The Future Lives Here: Guildford as a 21st Century Transit City
A Visioning by the Class of 2015 UBC Master of Urban Design

A project of

the School of Architecture and Landscape Architecture of the University of British Columbia

SALA

Funding partners:

CITY OF SURREY
the future lives here.

Published by the School of Architecture and Landscape Architecture and the School of Community and Regional Planning

Room 370 - 2357 Main Mall Vancouver, British Columbia
Canada V6T 1Z4

patrick.condon@sala.ubc.ca and scot.hein@ubc.ca

Editing by Korbin DaSilva, Manali Yadav, and Kristen Harrison

Graphic layout and production administration by Korbin daSilva and Manali Yadav

Contributions by Nastaran E. Beigi, Manali Yadav, Simone Levy, Amal Wasfi, Haneen Abdulsamad, Xinyun Li, Prachi Doshi, Lilian Zhang, Chen Fan, Hedieh Rashidi, Avishan Aghazadeh, Siyuan Zhao, Wei-cen Wang, Xueqi Wu, Yashas Hegde, Maryam Mahvash

Based on research and work resulting from the Winter 2015 Master's of Urban Design Studio, School of Architecture and Landscape Architecture

Includes bibliographical references
ISBN: 978-0-9780966-5-6

ACKNOWLEDGEMENTS

We want to acknowledge the many people who jointly participated in the class. Obviously the most important participants are the students themselves: Haneen Abdul Samad, Avishan Aghazadeh, Nastaran E. Beigi, Prachi Doshi, Chen Fan, Yashas Hegde, Simone Levy, Xinyun (Chloe) Li, Maryam Mahvash, Hedieh Rashidimalekshah, Weicen (Kate) Wang, Xueqi Wu, Manali Yadav, Lilian Zhang, Siyuan Zhao and Amal Wasfi.

Also crucial were our outside guests who guided our students and made them cognizant of issues particular to Surrey as well as enduring principles of urban design. We'd like to thank: Michele Alborg, John Bass, Patrick Chan, Patrick Condon, Patrick Cotter, Joyce Drohan, Frank Ducote, Mari Fjuita, Preet Heer, Scot Hein, Jean Lamontagne, Derek Lee, Don Luymes, Doug McLeod, Matthew Roddis, Daniel Roehr, YarOn Stern, Leslie Van Duzer, Kathy Wardle and Jay Wollenberg.

The class was organized and taught by Adjunct Professor Scot Hein. He was supported by Graduate Teaching Assistant Korbin DaSilva. Finally we want to especially thank the City of Surrey who's support for this project, both in terms of the generous availability of staff, and their financial support they have provided our students, made this design experience possible.

TABLE OF CONTENTS

Back row left to right: Korbin Dasilva, Avishan Aghazadeh, Hedi Rashidi, Patrick Condon, Maryam Mahvash, Lilian Zhang, Simone Levy, Manali Yadav, Nastaran Beigi, Xueqi Wu, Chen Fan, Don Luymes, Scot Hein

Front row left to right: Siyuan Zhao, Amal Wasfi, Haneen Abdul Samad, Prachi Doshi, Yashas Hegde, Weicen Wang, Xinyun Li

5

You hold in your hands the product of a very intense investigation, an investigation that occurred during the winter of 2015. Sixteen students in the UBC Master of Urban Design program took thirteen weeks to imagine a future for the Guildford Community in the City of Surrey BC. That City sponsored this investigation, contributing financial and staff support. The project? Imagining the transformation of a city from an auto oriented suburb to a transit oriented regional center with distinctive communities. Surrey BC, located in the center of its urban region, is adding more people every year than any other city in BC, and will soon surpass the City of Vancouver in population. By every indicator this city is a vanguard. It attracts more young families, more immigrants, and has faster job growth than its sister cities by far. It was for these reasons that we were happy to provide an urban design vision for this growing city and Guildford, one organized around walking, biking and transit rather than the car.

Surrey has significance beyond just this region as well, giving it even more trenchancy as a case study to investigate. Many of the transformations this city is experiencing are being felt in cities not just here in BC, not just in North America, but around the world. Surrey is the destination for more new immigrants to Canada than any other city in the region. In this way it is an "Arrival City", in the words of Douglas Saunders, author of a book of the same name. People from disparate cultures, with disparate skill sets, of various ages, and from differing economic circumstances all arrive here looking for a better life. For most of them the city has helped them accomplish their dreams. In the process the city is a much different place. Former industrial areas are now a hotbed of south asian enterprise. Homes once built as single family structures for "traditional" nuclear families of a working father, a stay at home mom, and three children now house extended families with many more members, most engaged in some form of entrepreneurial activity. Non traditional forms of work, retail, living, and education are rapidly eroding what were once clear planning boundaries between residential areas, job centres, commercial areas and schools. What are we to make of such complexity? And how should it now be regulated, if at all?

Meanwhile this city that grew with the car is being choked by the same device. Citizens and their elected officials are clear that they can no longer grow in a way that assumes auto mobility is universal, and that the street system can grow endlessly to allow for this metastatic growth.

But how do you transform a huge city that grew entirely around the car, the cul-de-sac, the parking lot , and the gas station, into a community where kids can walk to school, parents can bike to work, and seniors can take comfortable and convenient transit to doctor's appointments. Here again the issues faced by Surrey have relevance far wider than to BC or even Canada. More than 70 percent of North Americans now live in areas built around the car and these communities continue to capture the majority of new population. If we cannot find a way to reduce auto dependence then all of Canada's hopes for sustainability will be lost. But while the end game is clear the path from here to there (and now to then) is less so. The City of Surrey has arrived at a unique consensus for how they might achieve this goal. Citizens, city staff, and elected officials are unanimous in their support for a city wide light rail system - a system that will be, in time, more extensive than that of most North American metropolitan regions.

Thus, the first question for this studio was clear: How does the city use their transit initiative to transform an auto city to a transit city, a city where biking walking and transit provide first a real alternative to the car and, in time, provide the dominant way to get around. A follow up question: How do re-imagine and trahnsorm recognizing that all of our assumptions about land use, cultural homogeneity, and family structure no longer hold true.

Surrey is the very crucible of change. Not just here in our region but as an emblem of the changes that are informing new living patterns throughout the globe. Everywhere cities are expanding, doubling and sometimes tripling in size within a generation. A global movement of people away from rural conditions to urban ones is changing our way of life and even our sense of what it means to be a Canadian, to be a citizen more generally, to find a fulfilling life. What happens in Surrey matters everywhere. We are all Surrey. We at the UBC Urban Design Program are very proud to have had a chance to try to imagine a sustainable future for this amazing place.

Overview of 2015 Winter Studio Academic Experience

The 2015 SALA Masters of Urban Design Winter Studio expanded on the cohort's collective and individual achievements from the Fall Studio that declared growth strategies for Surrey, British Columbia at the city/regional scale. At the conclusion of the Fall Studio, Surrey officials requested that the second Winter Studio investigate the Guildford/104 Avenue Corridor precinct given anticipated light rail transit investment. This 2 mile by 1 mile precinct is characterized by varying, and complex, urban structure, related built form/typologies, a large scale economically viable mall that has recently enjoyed substantive re-investment, an active small business community, distinguished open space and natural landscape/water systems amenity and certain housing affordability allowing entry into the Canadian/West Coast market. The Guildford Precinct is recognized as a "market entry portal" that distinguishes it as an "Arrival City".

Given such geographic scope, complexity and attributes of the requested study area, the Winter Studio cohort approached the urban design challenge by simulating a professional practice consultancy, which might normally take up to a year in real time, in 13 weeks. The "studio design team" moved quickly through early investigations starting with in depth geographic/contextual analysis for each of the eight distinct sub-areas at ½ mile square, followed by early intuitive "first take" urban design aspirations for each square sub-area towards an understanding of the respective systems potential and visual/character signature. The studio cohort then "programmed" the design challenge by researching applicable policy intent/aspirations with Surrey staff experts towards a deeper understanding of the urban design, cultural and socio-economic role/identity that would distinguish Guildford. Given the urban design potential of the 104 Avenue Corridor as an organizational transit oriented "spine", the studio then investigated the prevailing context's visual signature using music as a metaphor. A streetscape musical composition, expressive of the corridor's visual characteristics, was produced and played/recorded towards strengthening contextual awareness/acuity.

Building on the efforts of the Winter Studio's early weeks focusing on achieving a deep understanding of context and policy, the cohort began the process of "practicing design iteration". Two teams, working independently, and with officials from the City of Surrey, developed alternate urban design frameworks. Shared review within the studio, and with Surrey officials, academic and economic advisors, provoked deeper questions beyond built form and urban pattern, to reveal how urban design can be in service to established, and emerging, local community identity and self-reliance while achieving new capacities that respond to housing and economic development needs. Through practicing a rigorous process of design iteration at the Guildford precinct scale, the studio sought the delicate balance between the responsibility of Guildford as an Arrival City, and as a place of new promise with respect to quality of life, civic pride and increased amenity, all within the context of a growing region.

Chapter 1: First Impressions

"Ready Fire Aim!" Absent of in-depth analysis, local insight and awareness of more specific municipal intentions, the studio produced a collective "first take" immediately after visiting the site. Teams of two urban designers each focused on eight 1 mile x 1 mile urban "patches", established by the local network of arterials. In these patches, the effort collectively worked to record urban patterns, systems, character, typologies and features. Given these observations, each team highlighted challenges and companion design opportunities. The teams intentionally worked independently from those analyzing adjacent areas. The "tiling" of the 8 patches, drawn at a large scale, motivated a studio discussion about the edges, respective transitions, larger systems potential and how Guildford might be recognized for a distinctive identity.

Existing Context

Located on the north side of 104 Ave. between 144 St. and 148 St., the study area is a low-density residential neighbourhood dominated by single-family housing. Some multi-family housing, mostly situated on the edges such as those on 108 St. and 148 St., dominates the multi-family housing character. Two elementary schools, one secondary school, two day cares, a fire hall station and two churches have given this area some of the amenities a small neighborhood needs. However it lacks other amenities, such as small public open spaces or plazas, urban green spaces (pocket or linear parks), as well as some local corner stores. The majority of the existing residential frontages are about 20.0 m providing a sense of place in terms of human scale. Unlike the favorable residential frontages and lot sizes, the frontages on 104 Ave. and the school plots are indeed challenging issues. Apart from the built form, the suburban character of the area and its auto-dependency is evident by inappropriate sidewalks (scale, form, pavement, furniture, lighting).

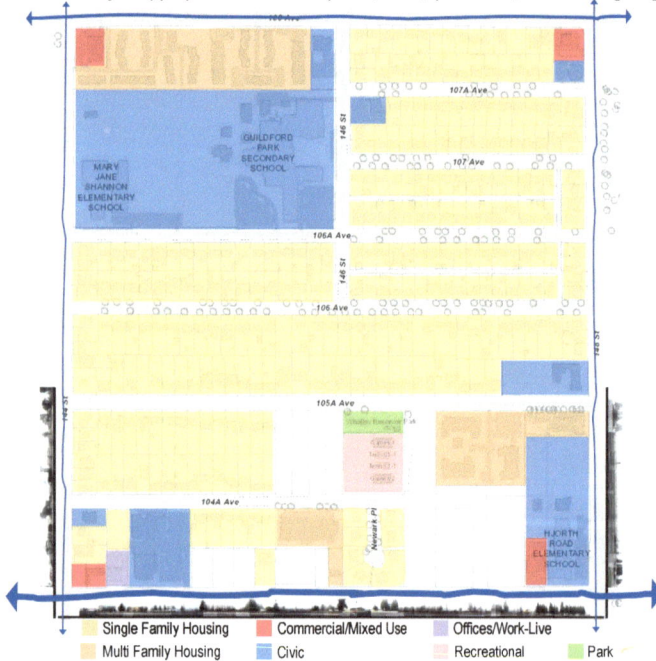

Opportunities and Challenges

The following major challenges are observed in the study area: lack of permeability both on the edges and within the area; lack of coherency of urban facade specifically on 104 Ave.; inadequate connectivity and accessibility; surface parking lots; building setbacks specifically on 104 Ave.; length of blocks; and the lack of small public open spaces accessible within a 400 m. walking distance. Despite major challenges, the area has potential to deal with some of the challenges. The following existing opportunities can be taken into consideration: existing parks on the both sides of the area and parks specimen trees; existing leftover lands and green lands; composition of built and unbuilt spaces; ditches and potential for bioswales; and the existing undeveloped lands on the southern edge of the area.

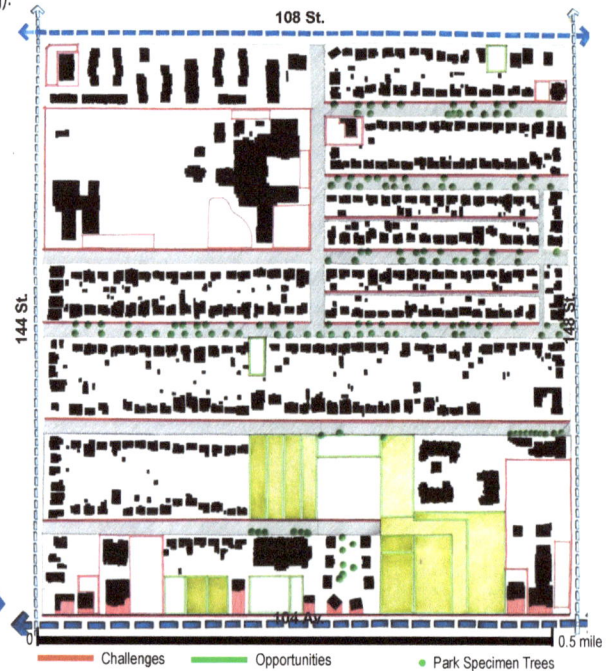

Figure 1: Existing land use. The diagram illustrates the domination of low-density residential buildings and the distribution of non-residential buildings such as schools, churches, gas stations, fire hall stations and day cares.

Legend: Single Family Housing | Commercial/Mixed Use | Offices/Work-Live | Multi Family Housing | Civic | Recreational | Park

Figure 3: Figure ground diagram combined with existing challenges and opportunities.

Legend: Challenges | Opportunities | Park Specimen Trees

Figure 2: Existing urban facade on 104 Av.

Team Members: Yashas Hedge & Maryam Mahvash

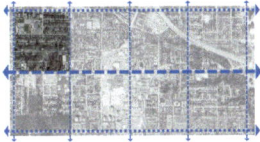

Produced Urban Framework

Along the edge of 104 Ave. most of the plots are subject to either redevelopment, new development or retrofit. Infill as a new strategy should be taken into consideration both along the edges and within the neighborhood. Surface parking lots and vacant plots are both eligible candidates for the infill strategy. The neighborhood can benefit from Hawthorne Park on the west side and Holly Park on the east side and has potential to be a perfect connector to complete a green network on a larger scale (Guildford/Surrey). Empty lands can serve as small public spaces or pathways to breakdown long blocks and provide permeability and better accessibility. The built and unbuilt plots offer an opportunity for adding lanes in some blocks in the area so that laneway houses may be implemented to gently increase density in the long term. Using the full capacity of exisiting plots on 104 Ave. in order to increase the density on the edge is highly recommended. Thus, the mixed-use strategy will offer vibrancy, vitality and safety to the edge.

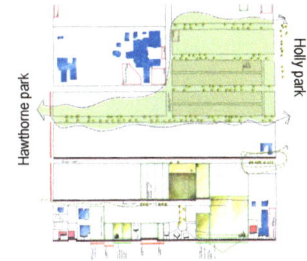

Figure 5: Neighborhood parks and existing parks could be connected as a green loop. This is a strategy ideal for both neighborhood and larger scale.

Figure 6: The demonstration of how future proposed mixed use could interact with the street by changing building policies. For instance, an extended roof overhang and the compound wall-building interface.

Figure 7: The sketch illustrates the urban pattern for green spaces which need to be developed on flat lands. Creating difference in levels with respect to the ground level. It defines a place making character for the green spaces.

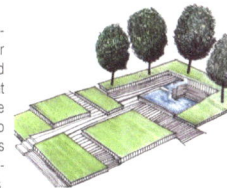

Figure 8: The parking lot in this sketch has a good scope to be converted into an urban plaza.

Figure 9: Schematic illustration which shows the urban pattern for this entire block is mainly governed by urban design policies.

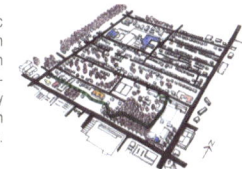

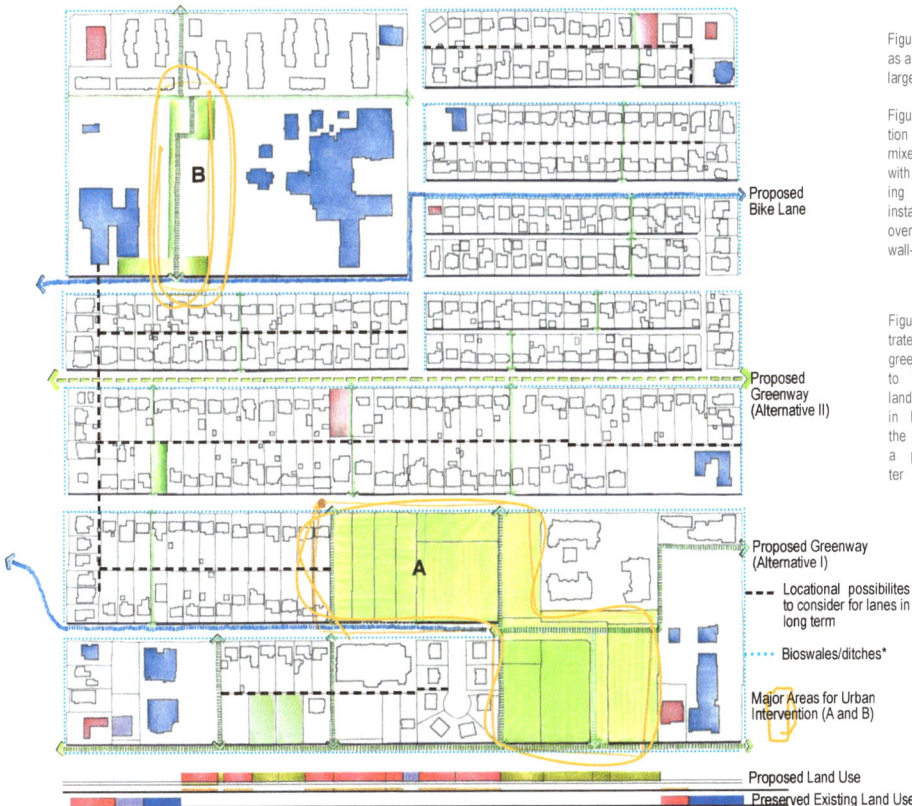

Proposed Bike Lane

Proposed Greenway (Alternative II)

Proposed Greenway (Alternative I)

- - - Locational possibilites to consider for lanes in long term

••••• Bioswales/ditches*

Major Areas for Urban Intervention (A and B)

Proposed Land Use
Preserved Existing Land Use

Figure 4: Diagram depicts the proposed major areas/features and proposed main attempts that need to be taken in to account for urban framework.
*Bioswales' locations in diagram are abstract.

References:
"COSMOS". cosmos.surrey.ca. City of Surrey. 2015. Web. January 2015.
"Guilford, City of Surrey". GoogleMaps. 2015. Web. 15 January 2015.

LEGEND: ⟷ OPPORTUNITY ⬭ OPPORTUNITY
⟷ MAJOR STREET ⟷ CHALLENGE ⬭ CHALLENGE

Existing Context

This site features a variety of uses. The prominent land use is multi-family housing in the form of various large apartment complexes. These are blocked off from each other and from the main streets by dense hedges and broken-off streets. The resulting condition is a landscape with very few through streets. One of the unique features of this site is the extension of the Guildford Town Centre in the form of a big box store connected by an indoor bridge within the mall. There are a number of surface parking lots along 104th Ave. surrounding the big box store and just north of this vast parking area is a large recreation centre and library for the community. An elementary school, small commercial strip, and large park near the northwest corner of the site complete the existing fabric.

Opportunities and Challenges

The new LRT line along 104th Ave. and the proposed stops at each end of this site present tremendous opportunity for change along the street. There is a chance to redevelop the large surface parking lots along the main arterial, and connect the recreation centre to the commercial hub of the Guildford Town Centre and the residential complexes on the site. There are a couple of run-down areas on this site, one of which has a few single-family homes in very poor condition. These places present opportunities for new, denser and more dynamic development to take place.

The main challenge of this site is the dominance of separated multi-family apartment complexes. These are important to the neighbourhood since they provide a more affordable option for families and their higher density is appropriate for the new transit corridor. The challenge is how to integrate them better into the neighbourhood fabric without compromising the buildings themselves. Better connected streets within the site would encourage people to walk from these housing areas to the arterial street.

Team Members: Simone Levy & Siyuan Zhao

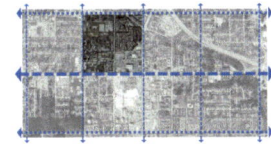

LEGEND:

Ⓣ	LRT STOP
⬅➡	LRT ROUTE
⬅➡	BUS ROUTE
⬅ ▭ ▸	MODIFIED LOCAL ROAD
(◖◗◖◗◖◗◖◗)	BIKE ROUTE
⬅--➡	GREENWAY(BIKES+PEDS)
◖●●●●●◗	PEDESTRAIN
⦀⦀⦀⦀⦀⦀⦀⦀	GREEN INFRASTRUCTURE
▬	MIXED-USE
▭	EXISTING CIVIC SPACE
▦	NEW OPEN SPACE
▦	EXISTING OPEN SPACE

Produced Urban Framework

In the design of this site, the corner of 104 Ave. and 152 St. has been re-imagined to include higher density mixed use development and public open spaces sheltered from the main arterial. These open spaces link the commercial centre to the recreation centre, and expand the community services of the area into the public realm. A smaller site of mixed use development is also proposed at the corner of 104th Ave. and 148th St., as well as new development at the site of poor quality single family houses, and intensification at the existing small commercial strip. This strategy provides interest points with commercial services within a short walk of every neighbourhood resident. In order to help people navigate the site better while preserving the current multi-family housing, local roads have been expanded wherever possible to better connect each complex to its surroundings. A greenway for bikes and pedestrians runs across the site along the existing park, and new protected bike lanes and sidewalks line the larger streets.

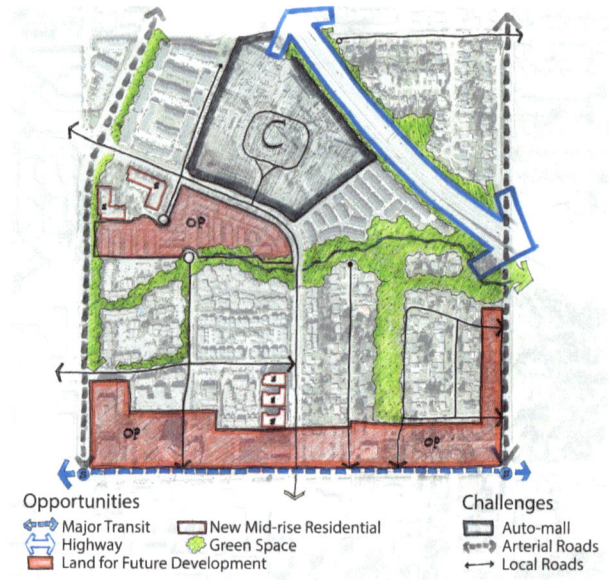

Opportunities
- 🔵 Major Transit   ☐ New Mid-rise Residential
- 🔵 Highway   🟢 Green Space
- 🟥 Land for Future Development

Challenges
- ☐ Auto-mall
- ⟵⟶ Arterial Roads
- ⟵⟶ Local Roads

Existing Context

First observations are primarily reflected based on the existing conditions through three main lenses: street pattern, building typology, and green system.

The site is bounded by three major arterials of 104 St., 152 St., and 156 St. These three corridors accommodate major public transit (buses) and daily traffic. When we walked further down the block, the neighbourhood fabric changed from corridors to a mainly dendritic street pattern.

The dominant building typologies along the main arterials are parking lots and vacant lands. Within the fabric, single-family housing and gated residential communities are the major typologies. In the northern part of the site lies a massive auto mall, which is separated from the rest of the neighbourhood by fencing on one side and the Trans-Canada Highway on the other side. In addition to typical suburban building characteristics on the site, new mixed use and mid-rise buildings have emerged.

There are isolated green areas on-site, but they are not connected or functional as a complete green network and public space.

Opportunities and Challenges

Opportunities
- Two LRT stops on 152 St. and 156 St. along 104 St.
- Bus route expansion through 152 St. and 156 St.
- Vacant land and parking lots along 104 St. for future development
- Potential to accommodate more population, commercial activities, employment and density
- Existing green spaces are natural resources with potential to enhance ecosystems, create a local public spaces, and link the missing bike and pedestrian pathways
- Popularity of local shops in the existing context (e.g. Korean Supermarket)

Challenges
- Lack of street connectivity with the neighbourhood fabric, dominantly culs-de-sac
- Transformation of big lots and single-family housing
- Neighbourhood segregation resulting from Highway 1 in the north side
- Gated residential community and lack of connectivity with the rest of the neighbourhood
- Lack of diversity and segregated location of auto mall

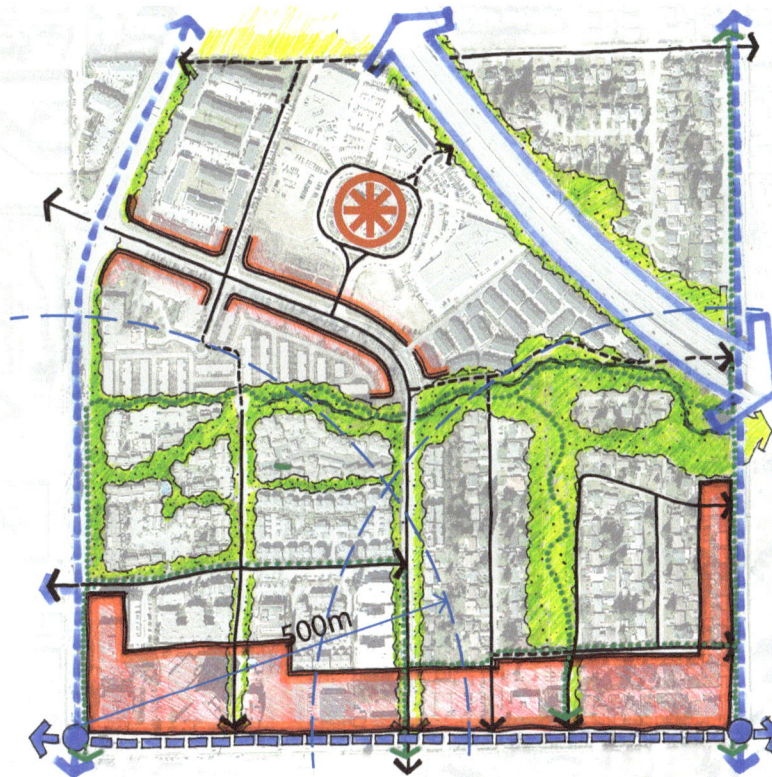

Team Members: Hedieh Rashidi & Lilian Zhang

Transit
⟺ Major Transit
⟷ Highway
⟺ Bus Routes
←→ Existing Local Roads
---- Proposed Sections of Local Roads

Building Typology
▬ Mixed-use Development Along Transit Corridor
▬ Retail Street Frontage
Automobile Dealership Hu

Green System
🌿 Green Public Space

Produced Urban Framework

Transit:
• Street connectivity is crucial in order to have better accessibility for the proposed transit system
• 5-minute walking distance to public transit
• Re-designing Highway 1 to a more neighbourhood-friendly corridor

Building typology:
• An improved streetscape for the overall image and character of Guildford area
• Mixed use buildings along the corridor to create a rich mixture of functions including walking, shopping and social activities
• Established local businesses with an international theme considering the various ethnic backgrounds in the area

• Transformation of auto mall into a hub of car dealerships to concentrated related business from other parts of the study area (Site 5 & 7)

Green System:
• Interconnected pedestrian and bike system
• Extended natural space to increase functionality, human activities and local characteristics

Fig.1: Study area - Block 4

Fig. 2: Existing challenges and opportunites for Block 4

Existing Context
A tale of two triangles

As we move away from the Guildford Town Centre, the streetscape changes gradually and the monotony of parking lots and big box stores is broken by the welcoming sight of a densely wooded area. The site is bounded by 104 Ave. and 108 Ave., and 156 St. and 160 St. Highway 1, which runs through the plot, divides the area into two triangles. While the entire area abounds with vacant plots, the triangle bounded by 104 Ave. and Highway 1 is densely vegetated and traversed by a serpentine creek that supports spawning salmon. The other triangle is punctuated with single-family homes and a few multi-family residences. Additionally, the site also boasts a school, churches, commercial buildings and a hotel.

Opportunities and Challenges

There is strong potential to integrate green vegetation into the area in order to improve health and safety in the neighbourhood. One measures that achieves such integration is preservation of the creek and its wildlife habitat. Given the close proximity of 104 Ave. LRT stop, re-designing the street frontage could accommodate mixed used developments and replace existing run-down homes. Big box stores could also be converted into mixed use buildings. However, the visual and physical disconnection that is created by Highway 1 proves to be a major challenge. In addition, the presence of vacant plots close to the highway adds to the area's challenges. It is foreseeable that single-family homes may undergo re-purposing in the future in order to double their population density and increase the efficiency of their built form.

Streetscape of block 4 is dominated by green silhoutte as opposed to built forms.

Team Members: Manali Yadav & Saki Wu

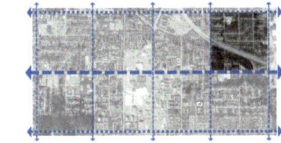

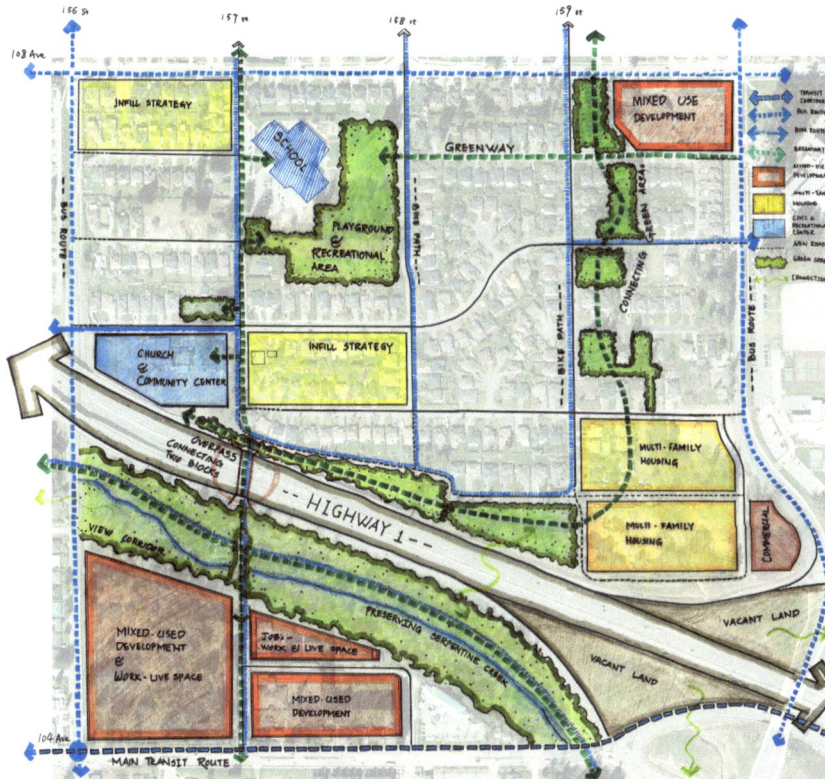

Fig. 3: Proposed land use, road and green infrastructure

Fig. 4: Proposed green connections

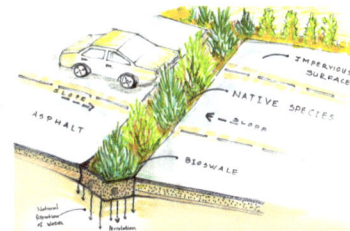

Fig. 5: Proposed bioswales for existing parking lots

Fig. 6: Proposed creek view with trails

Produced Urban Framework

Our vision is not just limited to the scheme depicted on Block 4, but also includes connections to adjacent blocks through the use of green corridors, trails, bike paths and visual connectivity (Figs. 3 & 4). By connecting the school, recreational parks and churches, these social and cultural spaces have potential to create an effective public realm. Accordingly, four key strategies that could be implemented in order to stimulate and densify the neighbourhood include: (1) In order to create a safer enviornment, the gap between two blocks should be bridged by refining the street infrastructure. A possible measure to achieve this is the addition of an overpass, which could act as a green corridor and emphasize an entry point. (2) In light of the unique topography of the area, drains could be designed to empty stormwater run-off into the stream. In the case of impervious areas such as asphalt-laden parking lots, bioswales (Fig. 5) are proposed. (3) New mixed used development close to the LRT stop could generate local employment, and buildings could be oriented in a manner that creates view corridors along the green buffer and reduces the sterile highway enviornment (Fig. 6). (4) Lastly, an infill strategy is proposed in order to create an equitable neighbourhood. These strategies can help Block 4 to seamlessly integrate with the adjoining blocks and create a more affordable & livable experience for new immigrants to the Guildford area.

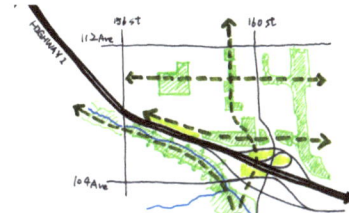

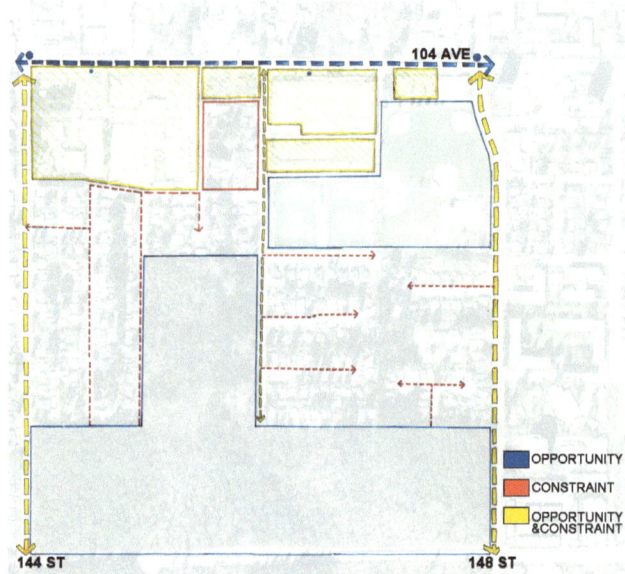

Existing Context

The site is located along the southern side of the 104 Ave., between 144 St. and 148 St. The site comprises a large urban forest in the south and a sports field in the northeast. The forest exudes a strong sense of nature but with limited accessibility and almost no human activity. The sports field with a large lawn creates openess. Along the 104 Ave., there are many parking lots. The largest one towards the northwest of the site belongs to an auto mall. There is a supermarket next to the 146 St. It has two stories and a parking lot behind it. It also has an underground parking lot. There are some small buildings dispersed within the site, including a Mcdonald's, two churches, a police station, and a bathware store. Other areas are mostly residential. There are single-family homes dominated with low building density. The overall building condition is poorly maintained. There are two existing bus stops on the 104 Ave. The traffic volume on the 104 Ave. is heavy. Walking along the 104 Ave. is uncomfortable because of the poor sidewalks.

Opportunities and Challenges

The map above comprises three categories. Blue, red, and yellow respectively illustrate opportunities, constraints, and the category both opportunity and constraint. In terms of opportunity, the urban forest is a potential place for people to explore nature and interact with wild-life. It is also important as an ecological preserve and contributes to the regional green infrastructure. The sports field provides recreational function for this and other neighbourhoods. Constraints include big block size, several culs-de-sac and weak street connectivity, which result in decreased walkability. Some areas have both constraints and opportunities. For instance, all surface parking lots are inefficient in terms of stormwater management, walkability and lack of land uses. However in some cases, their size and location provide potential opportunity for redevelopment. The supermarket has only commercial function within the building, so the building's value could be maximized by adding more mixed use above it. The arterial roads perform badly in terms of safety and comfort for the pedestrians, but they all have the potential to be refined under the new transit-oriented development agenda.

Team Member: Chen Fan& Kate Wang

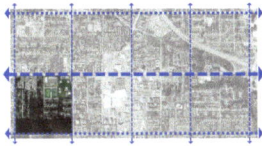

104 AVE

144 ST

148 ST

← - - →	MULTIWAY BOULEVARD
← - - →	PEDESTRIAN&BIKE
——	STREET NETWALK
·····	WALKING&BIKE PATH
▨	MIXED USE
▨	MULTIFAMILY HOUSE
▨	GREEN SPACE
⇕	GREEN CORRIDOR

Produced Urban Framework

Based on the analysis in last section, the urban design framework focuses on three parts.

(1) Street network: Black lines indicate the proposed street pattern. Most new blocks are walkable and form a grid pattern.

(2) Land use: Commercial and residential mixed use are proposed along the 104 Ave.

(3) Green system: A multi-functional urban park is proposed in the parking area to the south of the supermarket. It continues along the eastern side of the supermarket and connects to the existing small green space on 104 Ave. The sports field will be redesigned as a multi-functional space with an open field for playing variety of sports, rather than single use. The amount of work to be done inside the forest depends on further investigation. For instance, existing pattern of soil types, vegetation, topographical changes, and so on, could be evaluated at a later date. With more information, some trails, bike lanes, semi-open spaces, and enhanced habitat might be proposed for the forest. Some existing trees inside residential blocks will be refined as pocket parks, based on which, some residential buildings will be reorganized. All of these pocket parks, along with the urban forest and the recreational field will contribute to the regional green network.

Fig.1

Existing Context

Hovering over the above map, the grey area comes off as the most dominant and significant, the surface parking mostly catering to the Guildford
Town Center. Despite the fact that the mall is surrounded by this dry landscape, the mall itself is quite shiny, elegant and vibrant inside. Guildford
Town Center is the heart of the city; it acts as a hub and a collective point
for residents. This mall is an attractive spot for all surrounding neighbourhoods. The site extends from 148 St. to 152 St. from west to east, and
from 100 Ave. to 104 Ave. from south to north. It is a mix of residential and
commercial areas integrated into one another like two twisted L-shapes
forming an interesting fusion that could inspire future planning and designing (Fig.3). Although this site offers many services, it lacks civic facilities
such as schools, recreational areas, community centers and parks. The
street grid and the pedestrian grid are interrupted, and the green system
is not connected, presenting an unclear sense of direction. This site lacks
a musical façade where there is a gradual transition from one building
to another in terms of height and form. There are few vertically extruded
elements on the very flat area. The structural planning of the areas as
we move away from the mall slowly transitions from urban to suburban.

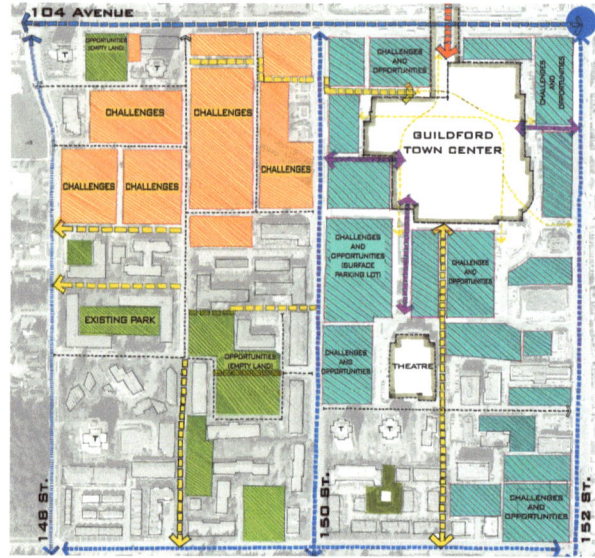
Fig.2

Opportunities and Challenges

Thinking about the area's planning vision for the future, most of the existing conditions are very challenging. However, there is great potential for
this area. Starting with the dominant grey area, the parking lots and some
empty private properties offer a lot of opportunities for future development
where one doesn't have to deal with any cultural and social constraints
(Fig. 4). It is mainly related to authoritarian decisions, unlike the residential
area on this site. The residential neighbourhood is mainly comprised of
single-family houses, where one has to deal with property owners, relo-
cating them or rearranging their own buildings and spaces, which is much
tougher and harder to deal with. They should be given alternatives where
they are safe and their businesses will not be negatively affected. Little
existing details could be inspiring to develop according to the community
needs. One such example is a barbershop with a residential façade. One
idea would be to develop a live-work building. The proposed LRT system
passing through 104 Ave. right next to the mall and having a major LRT
stop on the 104 Ave. and 152 St. intersection perfectly suits the location
of the mall. This helps create a homogeneously interconnected bus/LRT
system to easily accomplish the five minute walking distance concept.

23

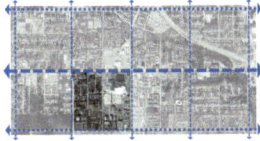

Team Members: Haneen Abdul Samad & Nastaran E.Beigi

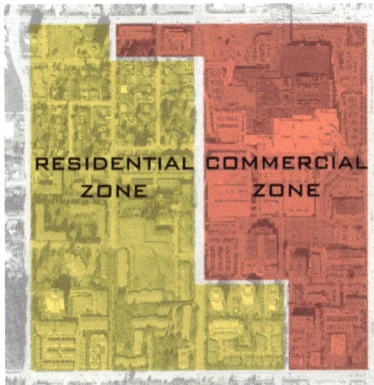

Fig.3

Fig.4

Fig. 4
The existing urban fabric illustrates the ratio between the built-up areas in black and the unbuilt areas in white. This demonstrates challenges and also the available opportunities for future developments. The amalgamation of super blocks as an urban pattern in this region has an adverse impact on walkability, safety and human scale.

Fig.5

Produced Urban Framework

A potential vision for this area starts with the mall itself. Since it is a hub and a prospective growth area for the whole region. As a first thought, the beauty of the mall from the inside is to be opened up and merged with its surroundings. All the internal walking patterns could be extended outdoors and integrated with the existing pedestrian/street grid to improve it. The majority of the parking lots are to be developed mainly into mixed use, mid-rise buildings. That will allow achieving a higher density in the area and a gradual integration with the existing context (the existing towers). Adding parks and greenways will help improve the ecological system, and adding streets will increase the efficiency of the street network.

LEGEND

■ Recreational	⋯ Bike lane
■ Residential	⋯ Bus lane
■ Mixed-use	▨ Opportunity and challenge
■ Commercial	▨ Challenges
■ Public open space	▨ Opportunity
→ Green ways	● Major transit
⋯ Ped. desire line	

Fig. 1: The existing context of the site

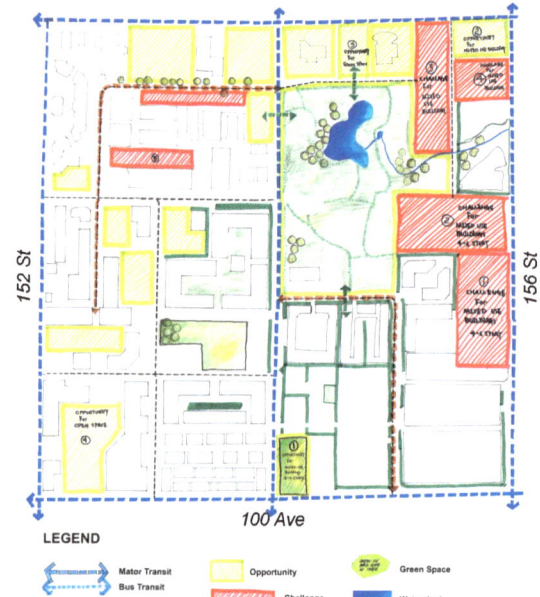
Fig. 2: The opportunities and challenges encountered on the site based on existing context

Existing Context

The area of Guildford is prominent today as a result of the proposed light rail transit (LRT) on 104 Ave. The half mile square enclosed from 152 St. to 156 St. and 104 Ave. to 100 Ave. is an amalgamation of two distinct building typologies. The Guildford Heights Lake bifurcates the residential area from the commercial domain. The Serpentine Creek merges with the lake flowing from east and crosses 156 St. Moreover the pride of the neighbourhood is the flag which is to the north of the site on 104 Ave. The Guildford Town Centre is across 152 St. to the west of the site. The area of the site near the Town Centre is predominantly commercial. However, further away high-density residential prevails. Crossing 156 St. to the east of the site is strictly single-family housing. This gradually diminishing density and zoning represents that the mall itself is the heart of the area of Guildford and the surrounding sites are the potential locations for future development.

Opportunities and Challenges

The major challenge encountered in this part of the region is density. Considering 104 Ave. this could be a thriving area with more jobs and residences. Therefore increasing density is of prime importance. Another point of concern is the connectivity between the buildings. The western portion of the site is comprised of offices and commercial strip malls that do not exhibit interrelation and all the buildings stand out alone. Not only does this make every building self-sufficient but it also leads to poor road network. Diversity in ethnicity is what makes Surrey unique, but the site does not encourage any cultural activity. This should be encouraged as the site holds a great potential for enhancing development surrounding the Town Centre area.

Team Members: Prachi Doshi & Xinyun Li

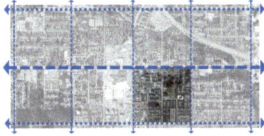

Fig. 3: The design strategies and proposed urban fabric for the half mile square

LEGEND

Mator Transit	Commercial Land Use	Watershed
Bus Transit	High Density Residencial Land Use	Green space
Bike Transit	Low Density Residencial Land Use	Public Open Space
Greenway	Mixed Use Land Use	
Pedestrian		

Fig. 4: (left top) represents the existing urban fabric. The white spaces indicate the parking lots. The physical space created here

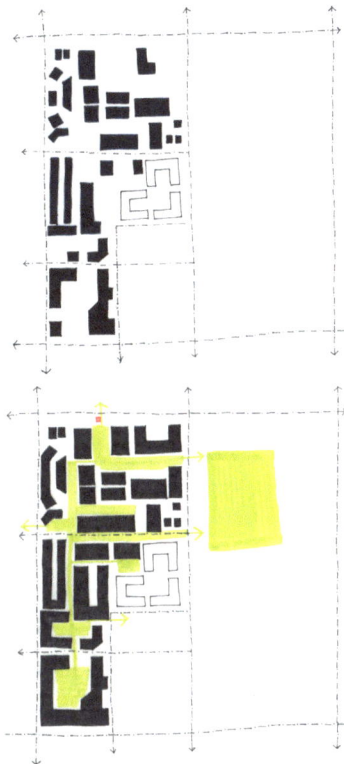

Fig. 5: (left bottom) represents the proposed urban fabric. The green path connects the urban plazas to the green park. The red square represents the canadian flag. It represents the filling of the empty spaces and connecting the buildings with a pedestrian path. East - West connection has a car access, but the North - South connection is entirely pedestrian.

Produced Urban Framework

A major feature of the concept reacting to existing conditions is the unveiling of the business park to the east of the mall. The huge parking lots segregate the commercial buildings. The newer concept is based on interconnectivity by eradicating the parking spaces. This provides an opportunity to use the surface as urban plazas. As a result, new commercial and mixed use buildings are proposed to fill the gaps. This derives a flow connecting all the urban plazas creating a pedestrian-friendly zone. Slow traffic is allowed to access 152 St., but a no-car-zone is created connecting the lower plaza to the lake via the flag. Newer buildings are proposed which are closer to each other and enhance the human scale pedestrian zone. Hence the functionality of the same physical space is altered again, providing better interconnectivity and more job opportunities. With this green zone terminating at the lake, trails have been proposed around the lake to encourage people to walk in the natural ambience providing a chance to plan for a recreational centre. The creek has been preserved and new zoning has been proposed to install mixed use buildings on 156 St., as the site adjacent to this is purely residential. Apparently this would lead to a sustainable development in terms of density, building typologies and the needs of humanity.

Existing Context

The northeastern side of the site is close to the intersection of 106 St. and the Trans-Canada Highway, which is connected to the Port Mann Bridge. Other major transit arterials surrounding this area are 104 Ave., and 156 St. A major feature of the site is Serpentine Creek, which is surrounded by a preserved green belt continuing from the eastern side of the site to the west. There are existing civic buildings and amenities on-site, including an elementary school and a hotel, both located along 104 Ave. The housing type for this area is mostly single-family detached, except some multi-family blocks close to 104 Ave. The neighbourhood consists of many dead end streets and culs-de-sac, making street connectivity an issue for this area.

Opportunities and Challenges

Opportunities:

1-Some of the existing parking lots and old buildings along 104 Ave. provide the opportunity for future development.

2-Opportunity for creating a connected street system within the neighbourhood, in order to promote walkability.

3-Opportunity for an interconnected public green space through walking trails within the neighbourhood and also along the Serpentine Creek.

Challenges:

1-Lack of connectivity within the street system, caused by dead ends and culs-de-sac.

2-Lack of a centralized commercial zone within the five minute walking distance of some areas within the neighbourhood.

3-Lack of connectivity between the open green space and the residential areas within the neighbourhood.

4-Lack of connectivity and communication between the buildings along 104 Ave. and the street.

Team Members: Avishan Aghazadeh & Amal Wasfi

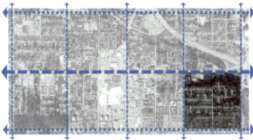

MAJOR
TRANSIT

GREENWAY

BUS

PEDESTRIAN

BIKE

PUBLIC
OPEN
SPACE

GROUND
ORIENTED
EMPHASIS

Produced Urban Framework

In order to promote walkability, a commercial zone is suggested around the intersection of 158 St. and 101 St., which is accessible within the five minute walking distance of the neighbourhood. Furthermore, street connections have been made through some culs-de-sac in order to create street connectivity as well as walkable paths. Trails have been considered along the Serpentine Creek and within the green belt. Green spaces within the neighbourhood will be connected through green pathways. The connection of the green spaces also promote walkability within the neighbourhood and also along the designated trails along the Serpentine creek. Bike routes have been suggested along the green systems with connections to the adjacent site. Since the location of the site is close to a main traffic intersection, this area can be considered as the entrance and gateway to the Guildford region. Therefore, a bridge is suggested in order to connect this site with the site across the road which will promote walkability and a connected green loop.

Chapter 2: Streetscape music

The organizing spine of Guildford is the 104 Avenue Corridor that anticipates a new light rail system. This two mile stretch anticipates new development opportunity for the entire length that takes into account varying existing conditions and context. Building on the studio's first impressions, the designers explored the corridor's existing and future identity using musical composition as a visual metaphor. The teams produced large scale overlay drawings of the corridor's north and south side streetscapes that visually emphasized the contextual rhythm, dynamics, textures, layers and proportional qualities. These drawings, or "contextual sheet music", were then played as a sonic composition to emphasize the importance of experiencing, or feeling, urban design character. This intuitive insight underpinned design explorations to follow.

Maryam Mahvash   Simone Levy

Yashas Hegde   Siyuan Zhao

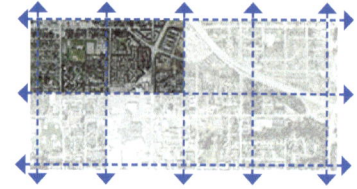

Musical Overlay

Street Scape

As we walked along the 104 Ave. corridor intersected by the streets 144 and 148 respectevely, we photographed the north side of the street to document the existing conditions. It is evident that a vast expanse of this urban fabric along 104 Avenue is undeveloped, resulting in poorly-maintained, vacant green spaces. The amenities on this stretch include a gas station, commercial units, seniors' rental housing, a church and two bus stops.The musical overlay clealy illustrates the monotony with respect to building heights and the vast expanse of open spaces.

The north side of 104 Ave. has a steady backdrop of low-rise apartment buildings between 148 St. and 150 St. The next stretch contrasts this, as the streetscape is mostly empty between 150 St. and 152 St. due to the large surface parking lots surrounding the extension of the Guildford Mall. Vegetation provides articulation in front of the monotonous buildings that line the street. When the visual identity of the streetscape is translated into music, the consistent, low-rise nature of the architecture is emphasized, as are the voids created by parking areas. The vegetation becomes extremely important, as it provides variation and rhythm at times when the built form disappears or remains constant at one height.

Lilian Zhang

Hedi Rashidi

Saki Wu

Manali Yadav

Fig. 1: Undulations of the terrain are best captured by Gheorghe Zamfir's El Condor Pasa

The dominant existing typologies along 104 Ave is mostly low-density structures, such as one-storey buildings, small-scale retail, and old single-family houses. Higher density towers, mid-rise, and mixed-use buildings have newly emerged on the site, but the neighborhood still embraces the suburban big-box supermarket, like the popular Korean grocery store. The significant landmark of the corridor is 20-storey Sheraton Hotel located close to 152 St. This site exhibits a transition from a suburban landscape to a more rural one for any pedestrians walking from 152 St. to 156 St. It is a transition from denser commercial areas (152 St) to single family homes and local retail (154 St), then towards dense trees and street-side gardens at the end of the section (156 St).

Like a musical score, while a city's topography can undulate over space and time. The preservation of the relationship between its elements ensures that its identity stays consistent. In this vision, block sizes, silhouttes, greenscapes and architectural articulations are akin to notes that lend a musical composition its distinct rhythm. For instance, Block 4 of 104 Ave. is brimming with greenery and even boasts a running stream. These harmonious features give the area the feel of a classical piece. On the other hand, the rundown appearence of tiny homes, broken sidewalk edges and the coarse asphalted road interupt the classical melody with loud, unpleasant notes (Fig.1).

2 STREETSCAPE MUSIC

Weicen Kate Wang   Haneen Abdul Samad

Chen Fan   Nastaran E.Beigi

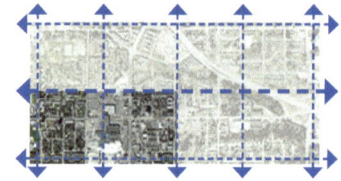

There are nine buildings in this area. All of them are one- or two-storey buildings. The section illustrates that two buildings are the widest, with the remaining buildings in the middle being small in footprint and close to each other. When these architectural features are translated into musical language, the low building heights are translated into only two musical notes (E and G). Some notes are closer to each other while some are further, which as a whole makes this section's music very irregular. In short, this façade can be represented with an intermittent and arrhythmic musical pattern.

The façade stretches between 148 St. and 152 St. on the south side streetscape of 104 Ave. It reflects a very non-rhythmic pattern, highlighting two contrasting facts. On one hand, the melodic line is dull for a lengthy portion of the façade. On the other hand we experience two shocking elements characterized by two 21-storey towers. Each tower extrudes from a very flat area mostly dominated by two-storey buildings. The soundscape has an unbalanced mode, which lacks hierarchy when transitioning from one state to the other. This is clear in the sketch below which reflects the physical shape as musical notes.

Observations

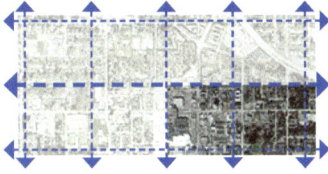

Prachi Doshi   Avishan Aghazadeh

Xinyun Li   Amal Wasfi

The streetscape between 152 St. and 156 St. is a transition between the Guildford Town Centre and the gateway to the neighbourhood. The tune of this changeover clearly depicts the abating tempo in terms of building frontage and increment in terms of vegetation. This waning is interrupted by the elevation of the tall Canadian flag, which is the only source of alteration within the existing monotonous streetscape. The soundscape clarifies that the trees yield a major sense of change across the street, while passing by the 104 Ave. The designer needs to consider the diminishing skyline at the horizon.

The dominant elements in this section are the trees. This area contains a belt of green space that is continuous throughout the block. At times the trees form a background to the buildings and they provide a rhythm that plays faster as we get closer to the end of this block. Series of buildings not higher than three stories interfere with this rhythm in some places. However, in general, since the buildings are at the same scale as the trees, they do not interrupt the rhythm.

Street Scape

Musical Overlay

Chapter 3: Policy and Programme

The studio's focus was to simulate a compressed, professional quality consultancy in service to our "client", The City of Surrey. As a result, investigative teams formed to liaise directly with municipal staff to ensure that deeper urban framework investigations reflected policy, and emerging best practice, intentions. Specific research was gathered around anticipated growth, demographic trends, related socio-cultural considerations, economic growth appreciating the "Arrival City" role that Guildford could continue to play, housing opportunity noting the importance of Guildford's current affordability, open space and related connectivity, food systems, water management and energy. These investigations culminated in a conversation with municipal representatives towards focusing the studio's design program while better understanding related opportunities for integrated strategies.

Legend:

- Urban Forest

- Parks & Recreational Area

- Vacant Land

- Creek

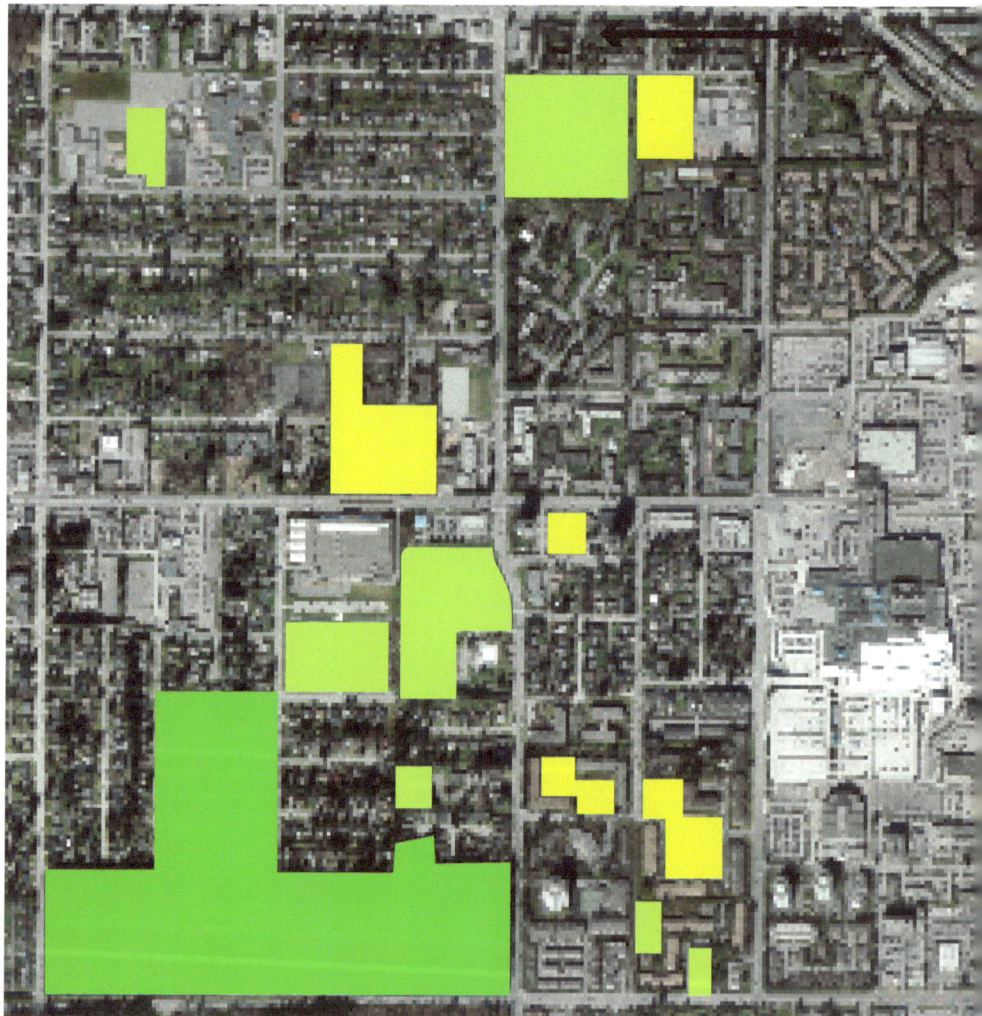

What We Have:

There are three types of existing green space. The first is urban forest located towards the northeast side and the other is to the southwest side of the study area. These natural assets will potentially retain regional ecology. The second type is recreational land that includes sports fields, parks, school playgrounds, etc. The third type is unutilized green space, such as some natural grassland without any use and maintenance. From researching the green system guidelines, there are some important issues that require extra attention: 1) the existing distribution of green spaces in terms of accessibility within 5 min. walk (400 m. distance) is a challenging issue, 2) the existing surface parking areas and the existing open spaces of schools and churches offer opportunity to meet the requirement for easy accessibility. Considering these issues, the design of small scale green and public open spaces will be taken into consideration.

Team Members

Saki Wu

Chen Fan

Nastaran Beigi

Maryam Mahvash

What We Need:

The lack of amenities in existing parks, including benches, playgrounds, pavement and lighting, is challenging. In fact, urbanizing the existing parks and open spaces in general needs to be taken into account. Besides, green lanes will provide a better accessibility and connectivity within the neighbourhoods while easing the implementation of a gentle infill policy within the blocks. Connecting existing parks and natural green infrastructure with greenways and a series of small open spaces, such as pocket parks, will complete both the green loop within the area and the green infrastructure network within the Guildford region. The overall strategy is to connect all green pieces together into a systematic green network. New parks, greenways or ecosystems will be proposed as corridors between any two green spaces that miss connections. This new green network will maximize wildlife habit, natural processes and human-nature interactions for the region.

Legend:

- **●** - Recreational & Libraries
- **●** - Schools
- **●** - Churches
- **●** - Senior Housing & Daycare

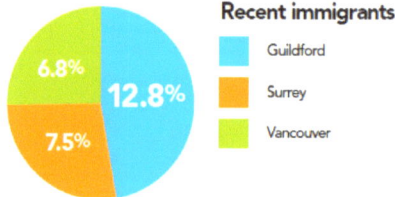

Recent immigrants
- Guildford — 12.8%
- Surrey — 7.5%
- Vancouver — 6.8%

Fig. 1: Recent immigrants comprise 12.8% of the total population in Guildford compared with 7.5% and 6.8% in Surrey & Vancouver, respectively1.

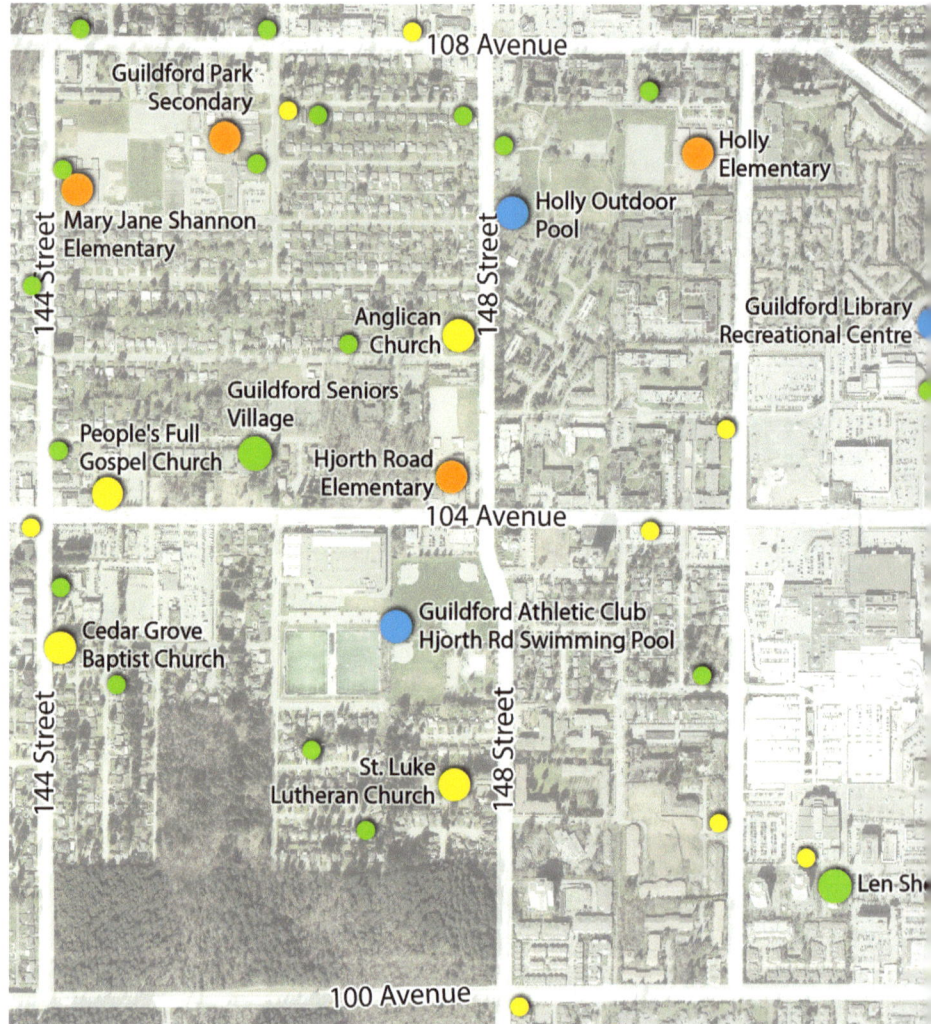

Map labels: 108 Avenue, Guildford Park Secondary, Holly Elementary, Holly Outdoor Pool, Mary Jane Shannon Elementary, 144 Street, 148 Street, Guildford Library Recreational Centre, Anglican Church, Guildford Seniors Village, People's Full Gospel Church, Hjorth Road Elementary, 104 Avenue, Cedar Grove Baptist Church, Guildford Athletic Club Hjorth Rd Swimming Pool, 144 Street, 148 Street, St. Luke Lutheran Church, Len Sh…, 100 Avenue

What We Have:

Guldiford, one of the six Town Centres in the City of Surrey, is often dubbed as an 'Arrival City' for immigrants. Guildford's population has grown significantly over the past decade, largely due to the influx of Asian immigrants. These events have greatly enhanced the diversity and vibrancy of the social and cultural fabric of Guildford (Fig.1). Notably, immigration has directly contributed to the rise in the number of churches, schools, community centers, recreational centers, libraries, day-care facilities and senior homes in the area. In addition to quantity, even the quality of these social spaces has improved significantly. For instance, the Guildford Recreational Center and Library is a state-of-art facility that organizes programs for seniors and youth. Moreover, as highlighted in the four leftmost blocks in the map above, social amenities and facilities are judiciously distributed throughout the area. However, their concentration exhibits a marked decrease as one moves away from the Town Centre. As a result, residents rely heavily on private transport and the sparse population diminishes the social realm.

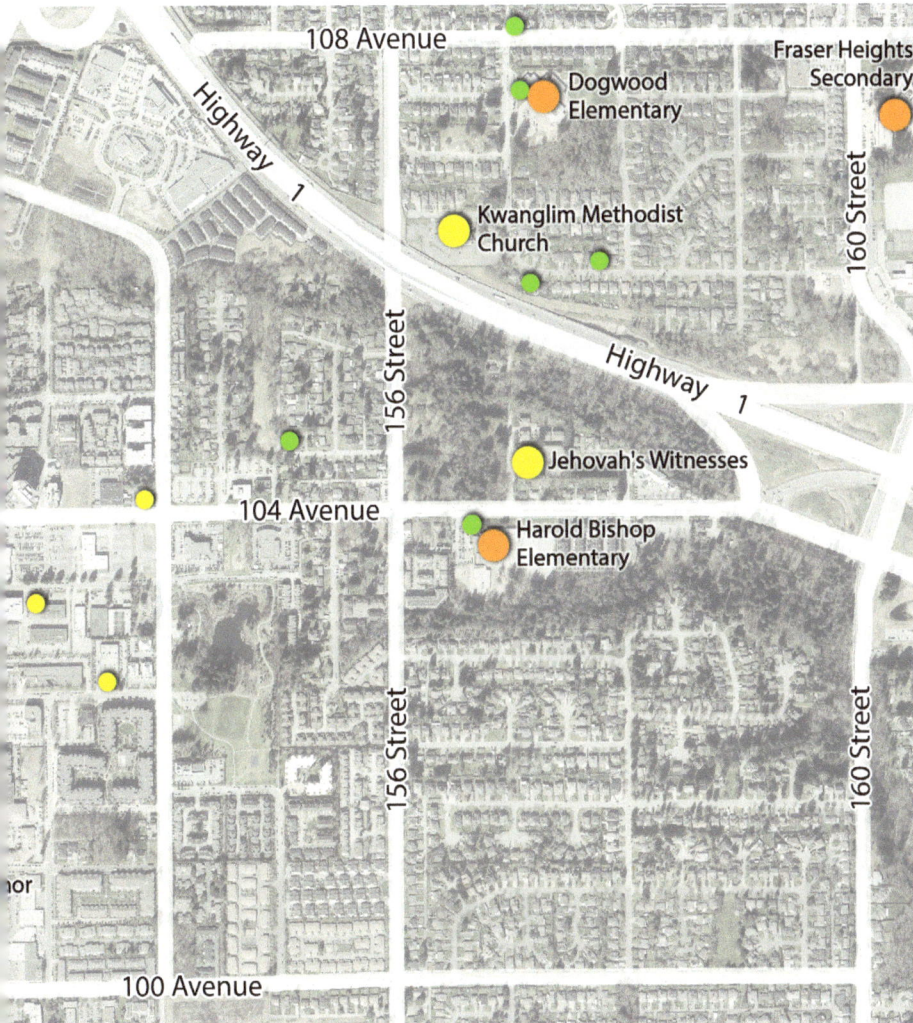

Team Members

Manali Yadav

Weiceng Wang

Siyuan Zhao

Lilian Zhang

References:
1. United Way Lower Mainland. My Neighbourhood My Future: Guildford West, Surrey. 2011. http://bit.ly/1F8HsUq

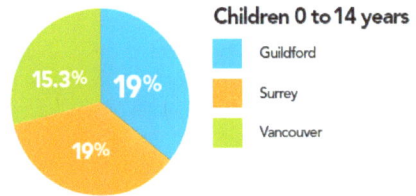

Children 0 to 14 years

- Guildford
- Surrey
- Vancouver

Fig. 2: Childern upto 14 years of age account for 19% of the population in Guildford. In Surrey and Vancouver, these numbers are 19% and 15.3%, respectively1.

What We Need:

In order to create a more inclusive and equitable neighbourhood for all residents, it is important to identify missing nodes within the existing fabric. In light of population growth, providing more facilites such as neighbourhood housing, day-care, senior housing and community centers in the more distant parts of Guildford could foster greater engagment within all the ethinc and age groups. Additional emphasis must also be given to ensuring equitable access to youth programs for families from all economic strata (Fig.2). In order to build a safer and more secure community, these social centers must be connected by more efficient and sustainable means of transport such as pedestrian paths, bike lanes, trails and greenways. Furthermore, the regular occurrence of events such as farmers markets, community gardening, fairs, open-air amphitheater shows and outdoor excursions will greatly animate the social fabric of the region. Promoting multi-cultural activities and events will foster a stronger community feeling within the neighbourhood.

Amal Wasfi

Cultural Facilities for the First Nations in Surrey:

What We Have:

Metro Vancouver's urban Aboriginal population is 40,000-60,000 people, the largest population in British Columbia, with the highest social needs of any population group [1]. Cultural-based social and health centers are critical to the well-being of the urban Aboriginal community. The Aboriginal believe in "Holism", the Physical, Emotional, Spiritual and Mental balance, with particular attention to congruence between the body and mind encompassed by the spirit. These Four Pillars together in combination make up a good overall well-being. It is important to mention that part of the spiritual activities is an attendance of a place called the "Sweat Lodge". It is a place to purify the body and soul. Expressing traditions and building faith is mandated to improve their quality of life . Currently there exist two hubs in North and West Vancouver

West Vancouver Community Hub

which allow for the adherence of these Four Pillars. However, in Surrey, there is only a park and a school on 104 Ave. and 130 St., but there is no community hub in the areas south of the Fraser River. Instead of an 'out of sight and out of mind' mentality which applies to much of the region's population, having these facilities in the Town Centre will help in bridging the gap and blending the Aboriginal with the different cultures.

What We Need:

It is recommended to provide a hub in the middle of the South Fraser region with easy access from Highway 1 and the Pattullo Bridge. Although the First Nations' population in Surrey is around 2,000 people, this proposal is looking to serve the Aboriginals in the South of the Fraser Area (pop. 33,515) living off reserves in dispersed groups [2]. The proposed center will have the four mentioned main facilities as followed: 1) exercising (Native sports) facilities; 2) "Aboriginal Circle of Life Services Program" facilities delivered by certified workers; 3) spiritual structures; and 4) education and art outreach facilities. The art facilities could be expanded to include art shows/sales, especially if the location is close to the "proposed" gateway visitors village (between 156 Ave and 160 Ave).

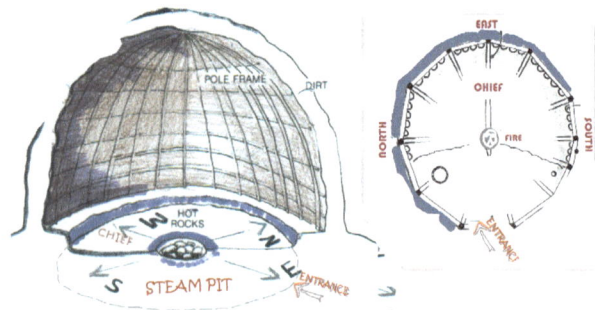
Sective and plan of a "Sweat Lodge" - a health and symbolic place to purify the body and the soul by sweating inside a heated tent.

References:
1-Metro Vancouver Aboriginal Executive Council Report, 2014. .http://www.mvaec.ca/infoshare
2-Profile of Aboriginal People in the Fraser Health Region, 2010, http://www.fraserhealth.ca/media/Aboriginal%20Profile_2010.pdf

Xinyun Li

Cultural Facilities in Surrey:

What We Have:

The existing cultural situation in Guildford can be better understood by looking at the past and reflecting on the present and trends for the future. The western culture that emerged in Vancouver region in the late 1800s was preceded by a Native culture that was rich in art, theater and dance (Bruce Macdonald,1992). Surrey is one of the most rapidly developed regions, with nearly 10,000 new residents added every year. Cultural facilities are significant to enhance urbanization and to create a sustainable, ethnic and culturally diverse and socially cohesive community. The Guildford area has several cultural facilities: Guildford Youth Lounge in Guildford Recreation Center, Guildford Islamic Cultural Center (Fig 1). These facilities promote public art exihibitions and religious activities. However, there are some issues to resolve. Most importantly, it lacks cultural facilities for different ethnic groups. Cultural facilities should be vibrant, muti-cultural and intimate places for all ages to enjoy. It should create a "sense of place". Also, lack of open space leads to less interaction amongst the residents and creates unsafe environment.

Fig.1: The Guildford Islamic Cultural Center. It should improve its cultural identity and also enhance characteristics of the exterior open space.
Sources: Google Earth

What We Need:

Recommended strategies:

1. Considering the different cultural and historical background of immigrants, the Guildford community should create multi-cultural facilities that promote community engagement. For example, public art galleries, museums, theaters, religious places.

2.Creating more public space and landscaped areas encourages people to communicate and perform more cultural activities. We can design cultural sculptures and green schemes for the existing open spaces and green spaces. For example, forest theatres and the cultural plazas (Fig 2).

3.Creating greenways that can form a continuous loop for the Guildford area. It would connect all the cultural facilities as one cultural loop. It will then provide people with an easily walkable, vibrant, and culturally rich walking experience.

Fig.2: The cultural scuptures and landscape archifacts reflects Surrey's culture combined with the greenway. It is a excellent example that represents Guildford's cultural identity.
Sources: Public Art Program in Surrey.

References:
1-Surrey Public Art Plan 2012-2016. Public Art Program.http://www.surrey.ca/culture-recreation/1654.aspx

Proposed and existing housing typology

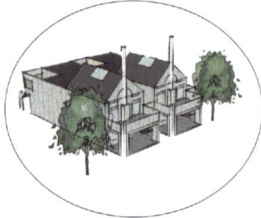

Fig 1. Infill Strategy for single-family housing

Fig 2. Proposed Strategy for new mid-rise developments

Fig 3. Existing high density towers

What We Have:

Although the dominant type of residence in this area is the single-family detached home, there is also a diversity of housing typology that includes mixed-use, multi-family residential, high-rise towers and mid-rise residential. The two ends of the site (northwestern and southeastern parts of Guildford) are mostly covered in single-family homes. As we go further towards the centre of Guildford, there is a variety of mid-rise residential units and multi-family housing. High-rise towers are mostly concentrated in the centre of the site and close to the Guildford Town Centre. There are many existing vacant lands and parking lots along the corridor and also throughout the fabric with the potential of accommodating future housing developments.

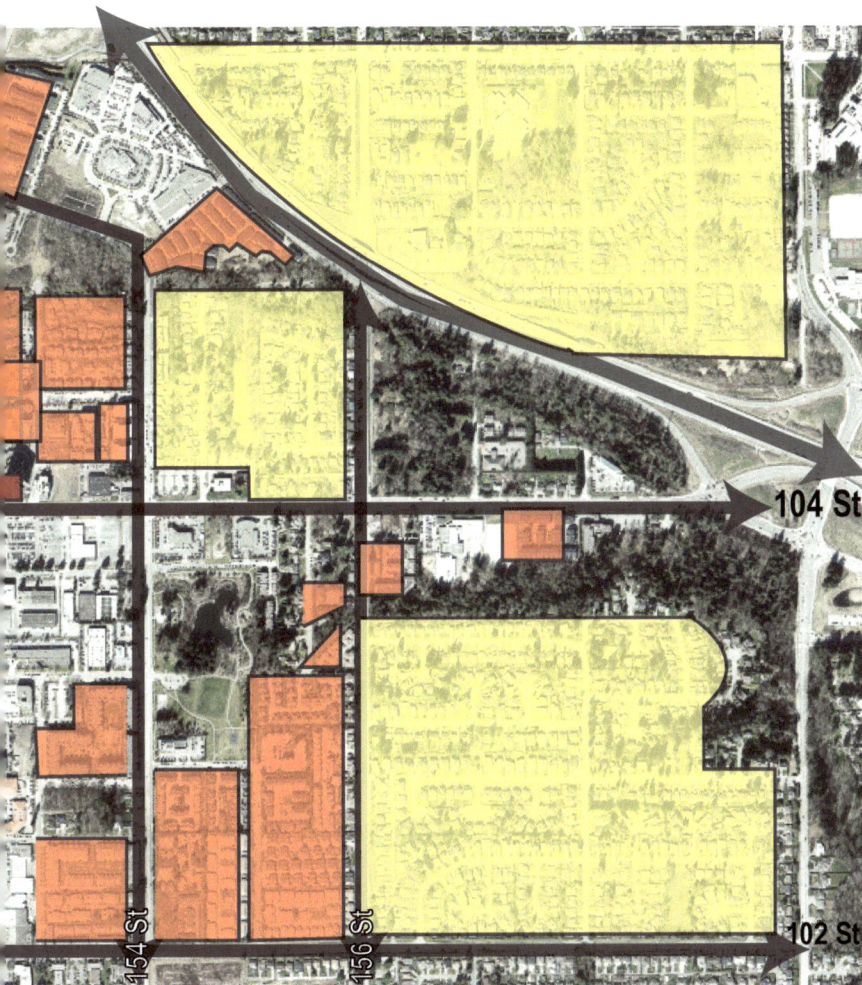

Team Members

Avishan Aghazadeh

Hedieh Rashidi

Legend:

Single-family Housing

Mid-rise Residential Housing

High-rise Residential Housing

Existing Affordable Housing

What We Need:

Due to the proposed LRT system along 104 Ave., an economic boost and additional population growth is expected within the span of the next 40 years. As a result of this transition, an increase in density and number of housing units is demanded in order to accommodate this population. A set of densification strategies are proposed with respect to the complex fabric of the site as following:

• Infill projects can be considered for the lower-density neighbourhoods (single family housing) by adding secondary suites, laneway housing, and row houses. (Fig 1)

• Development of 4 to 6-storey buildings for the potential areas (existing parking lots and vacant lands). An example of proposed housing typology could be similar to the mid-rise building at the north-western parts of 104 Ave. and 154 St. intersections. (Fig 2)

• The recently-built, high-density towers are expected to remain in the considered period for 40 years. (Fig 3)

• Older affordable residential buildings (shown in blue) can be redeveloped with respect to the existing footprint and the retention of the greenery.

Legend:

- **LRT Transit Stops**
- **LRT Transit Lanes**
- **Bus Transit Lanes**
- **Trans-Canada Hwy**
- **Truck Lanes**
- **Bike Lanes**
- **Existing Green Lanes**
- **Proposed Green Lanes**
- **Existing Internal Roads**
- **Proposed Internal Roads**
- **Cul-de-sac**

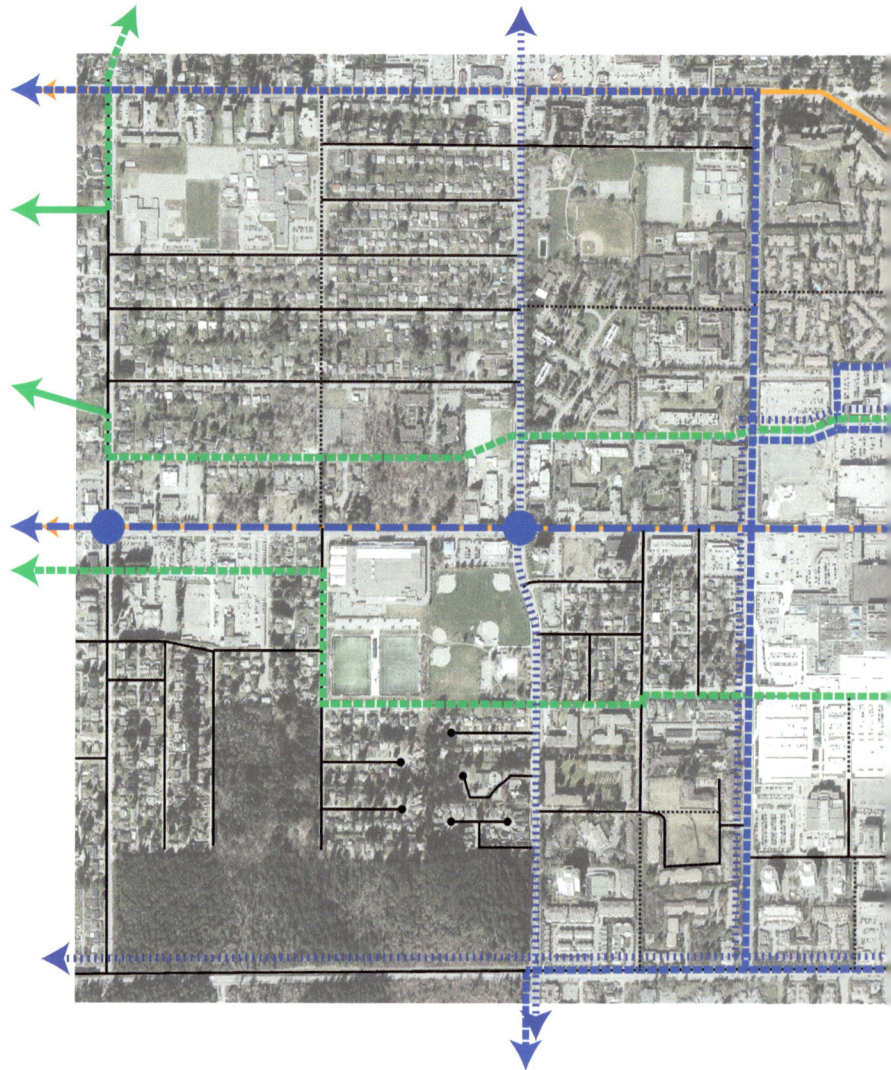

What We Have:

Transportation in Surrey is about finding smarter travel choices for people and interconnecting the neighbourhoods. Guildford is home to 104 Ave., which is a major east-west arterial connecting Surrey City Centre to the Trans-Canada Highway. It is predominantly a car-dominated neighbourhood, even though it exhibits a number of greenways and bike lanes. The bus network is one of the popular modes of public transportation. The presence of some greenery (forestry spaces), as well as the palimpsest of various old transit modes ameliorates the travel experience. The street grids surrounding the Guildford Town Centre are poorly determined. As we move away from the Town Centre, the blocks' structures adopt the cul-de-sac street format. Thus, we have an established neighbourhood suitable for alternative transit options.

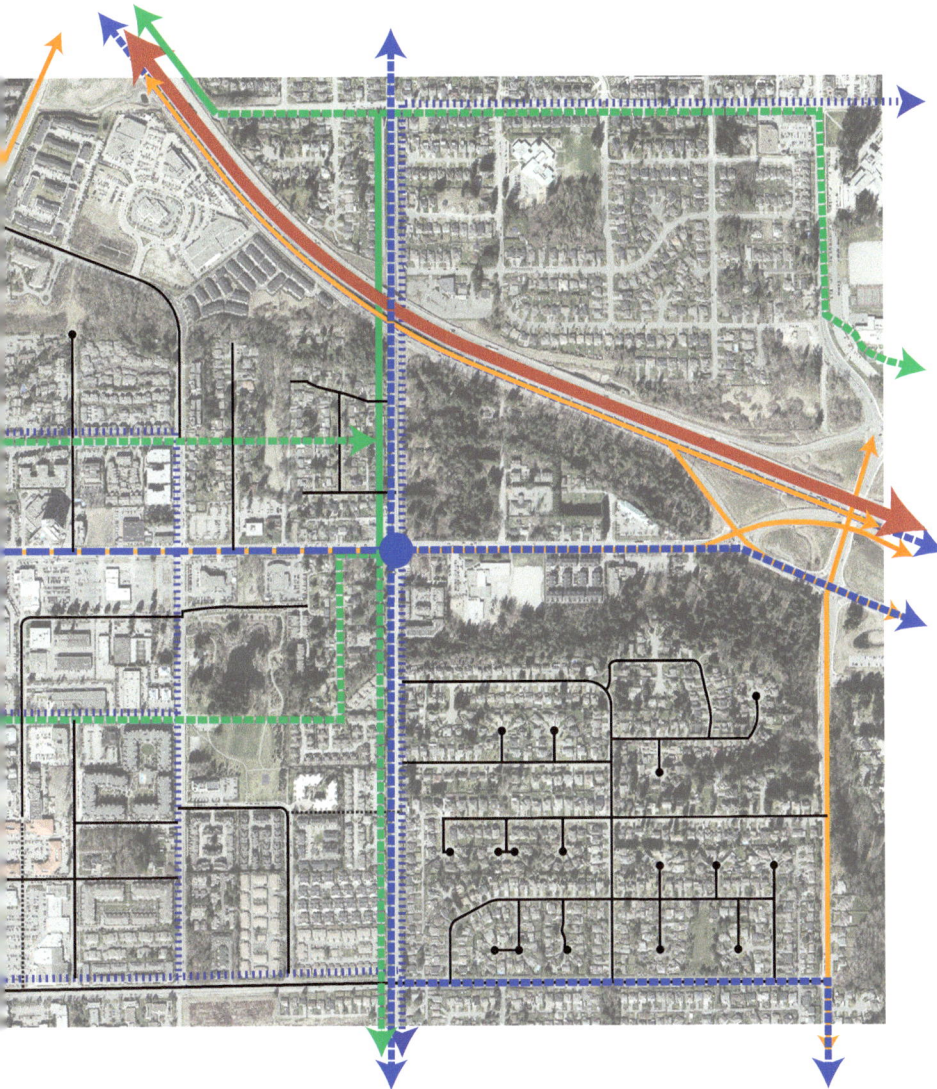

Team Members

Haneen Abdul Samad

Prachi Doshi

What We Need:

The advent of Light Rail Transit would give a boost to 104 Ave. in terms of density and accessibility. There would be a need to maximize the number of people accessing the transit nodes. Along with 104 Ave., 102A Ave. and 105 Ave. will be two potential streets that can absorb traffic. The existing transportation layout calls for resolving the grid pattern for local roads. Abruptly ending roads lead to weak connectivity. Guildford has bike lanes that need to be redesigned to form loops, which would encourage bike usage. The neighbourhood lacks pedestrian-friendly sidewalks. We foresee Guildford as a walkable and bike-friendly neighbourhood, which would not only be sustainable but would also enhance the street life.

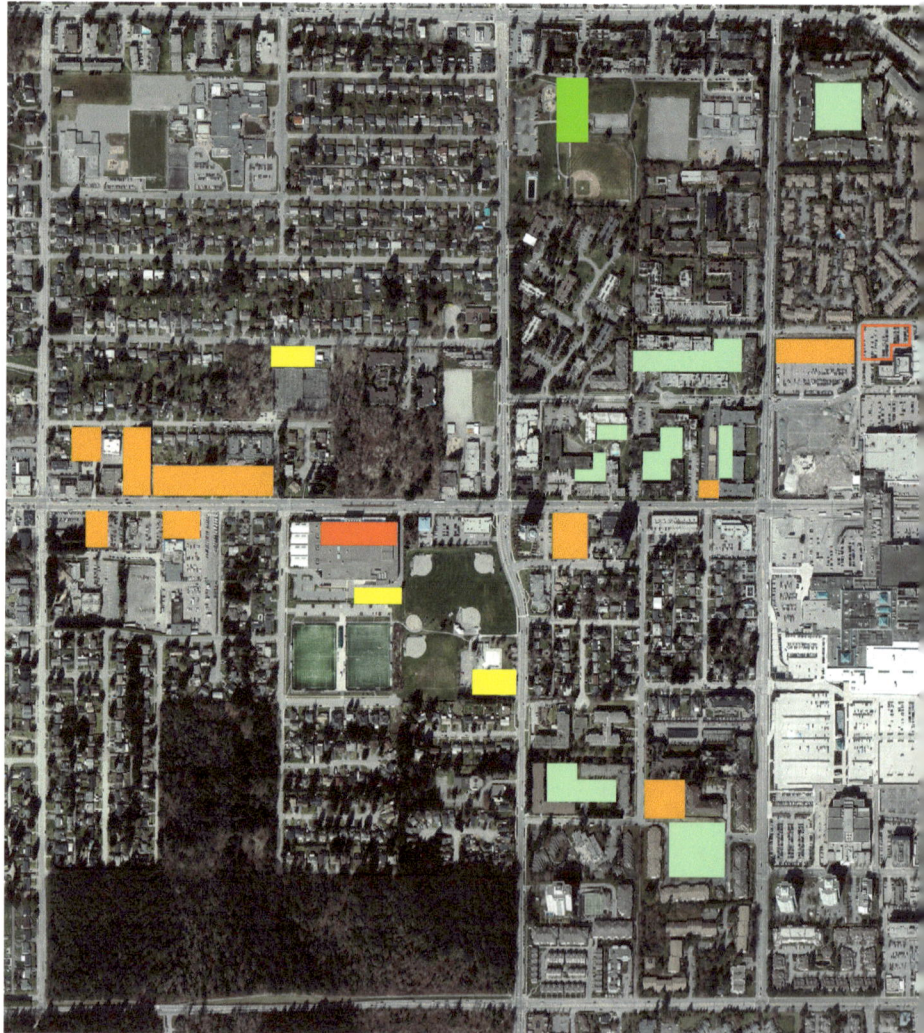

Legend:

- Existing Community Garden

- Potential Garden Sites Within Existing Park Lands

- Potential Agriculture Sites Within Private Open Lots

- Areas Within Private Developments That Could Incorporate Community-Scaled Agriculture Space

- Stores With Some Groceries

- Potential Farmers' Market Site

What We Have:

There is an existing community garden in Holly Park to the north within the Guildford study area. There is also an abundance of city parks throughout the site, as well as private- and publicly-owned vacant lots. There is better potential for temporary urban agriculture on private lots with the hope of integrating food production into future developments. Currently, there are several markets selling produce in Guildford, but most of these do not feature any locally harvested foods. The Guildford Recreation Centre and Library is a centrally located facility that could function as a site for a local farmers' market for the neighbourhood. Surrey currently has a food security plan with similarities to Vancouver's very successful strategy.

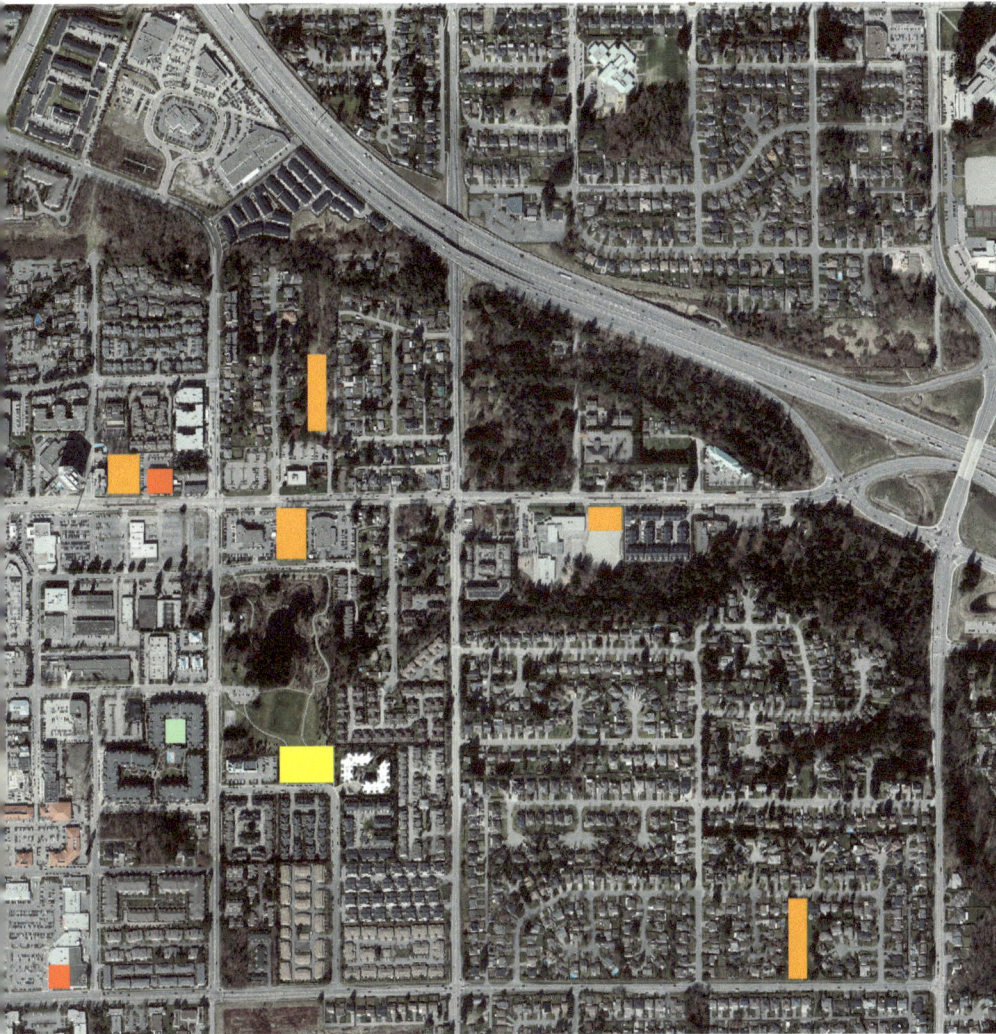

Simone Levy

What We Need:

Guildford needs a network of community gardens, urban farms and orchards. A strategy must also be implemented to make local produce readily available to all residents. This can be done through a 3-tier system: 1) a large farmers' market at a central, city-owned location, 2) produce markets within a 5 min. walk of all residents, and 3) stands located at urban farm sites. Community-oriented commercial kitchens can promote local food-related businesses. Large-scale integration of compost, backyard chickens, and pollinators can improve farming productivity. New developments should incorporate urban agriculture in the form of gardens and roof-top farms. Non-profits should be located within schools to run gardening programs.

Legend:

▬▬ - District energy lines

● - Geo Heat Exchangers

What We Have:

The diagram above illustrates the district energy system integrated with the Sewer Heat Recovery Plant. The Sewer Heat Recovery Plant is represented above to show its connection to the district energy system. However, it does not represent the location of the plant. The district energy lines are not extended to areas which are not under the scope for future development as it is not economically viable. Geo-heat exchangers are placed in areas of high energy flow ideally in a surrounding park or linked with the transit station.

Yashas Hegde

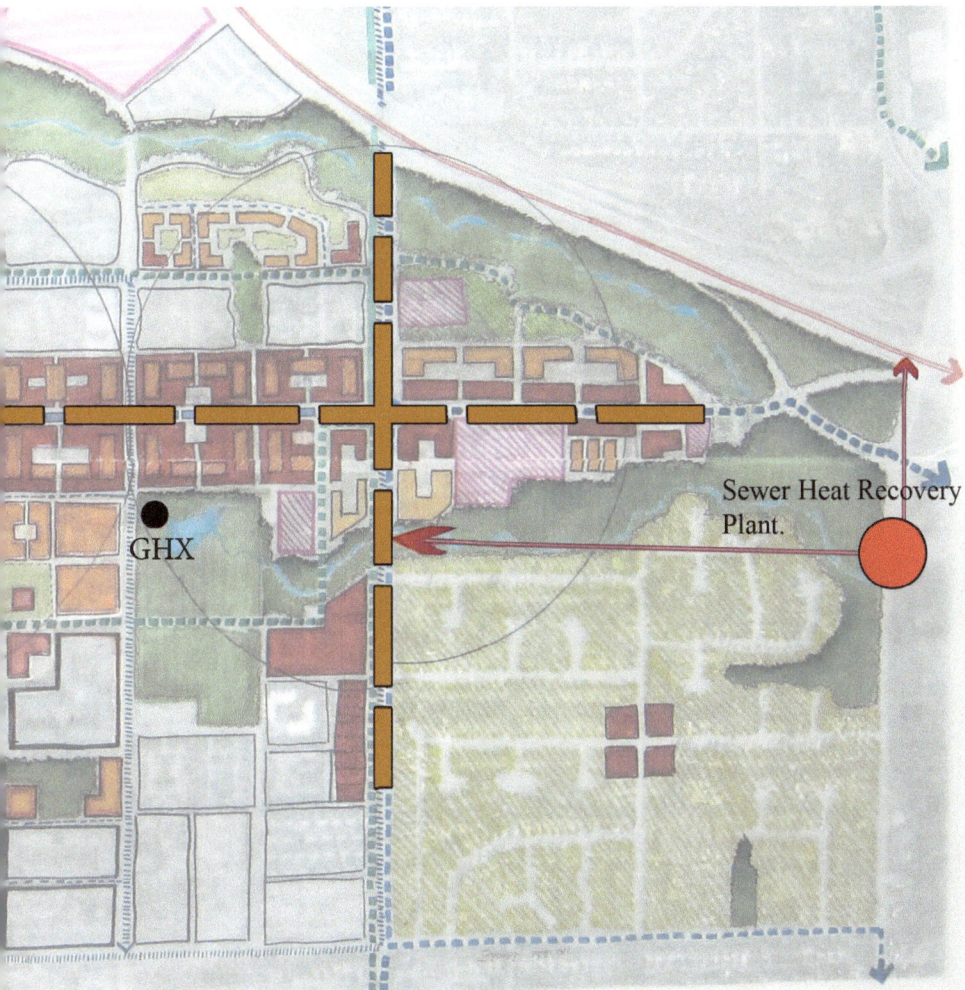

GHX

Sewer Heat Recovery Plant.

What We Need:

Based on location, the closest proposed Sewer Heat Recovery Plant is the Quibble Creek Pump Station, situated in 94A Avenue at King George Highway which could be a possible connector to the district energy system. The proposed district energy is based on the zoning probabilities and since the above zoning is subject to change the district energy system is also subject to change accordingly.

Chapter 4: Mall in? Mall out?

Located near the geographic center of the study area that anticipates a new transit stop, and arguably the cultural focus, the future of the Guildford Mall was acknowledged as a strategic driver of any future urban framework. Appreciating the recent substantial investment to improve the mall, while also noticing the rather introverted character of malls in general, the studio speculated on two alternative futures. This investigation was conducted as a "mini design charette" with the studio dividing into two teams to each pursue a future with or without the mall. Alternative scenarios, intended to be grounded in economic reality and pragmatic implementation, were generated and presented. A comparative conversation with senior planning leadership to conclude the charette provided greater strategic insight into, and assisted in "calibrating" the studio's shared values about, the important future of Guildford Mall.

Haneen Abdulsamad

Avishan Aghazadeh

Chen Fan

Simone Levy

Status Quo Medium Density Housing (Typ.)

Cultural Hub & Public Open Space

Gentle Infill Strategy (Typ.)

104 Ave. LRT Corridor

New Affordable Housing

Guildford Mall

WESTSIDE DESIGN

• High-density commercial buildings are introduced along 104 Ave. Building typologies include courtyard buildings on the northern side and podium-towers on the southern side that accommodate the relocated car dealerships from the auto-mall off Highway 1.

• On the northern side of Guildford Mall, a cultural hub is proposed with an open space for festivals and concerts. This site is surrounded by cultural and institutional buildings and an entertainment district.

• Green spaces along the corridor will be preserved in order to maintain a well-connected green network.

• The mall is preserved. Mixed-use buildings are proposed on existing parking lots. Some retail and light industrial spaces are reserved for start-ups south of the mall.

Chloe Li

Amal Wasfi

Maryam Mahvash

Siyuan Zhao

Repurposed Auto-Mall Site

Urban Agriculture

Entertainment District

Revitalized Serpentine Creek System

Existing School (Typ.)

Corner Store Location

Live-Work Units, and Small Business Start-ups

EASTSIDE DESIGN

- Mixed-use podium-towers line 104 Ave.
- Bars and small business are located on the eastern side of the mall.
- Lower density culs-de-sac to the south are opened in places to allow for better circulation. Some gentle density intensification takes place along with the addition of small corner stores.
- Green spaces are preserved and enhanced and pathways are proposed to connect them. This takes the form of a bridge at the eastern end, which also acts as a gateway.
- The existing auto-mall is repurposed as a trade school.

Haneen Abdulsamad

Avishan Aghazadeh

Chen Fan

Simone Levy

HOUSING STRATEGY

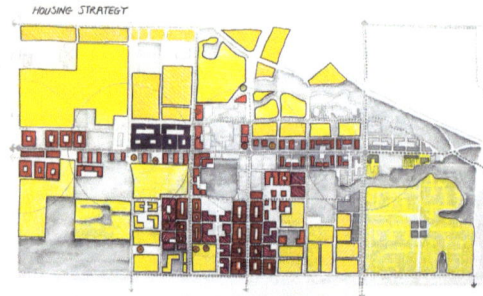

	Existing/Low Density		Affordable Housing
	Medium Density		Mixed-affordability Housing
	High Density		Live-work Units
	Market Housing		

The housing strategy aims to provide a variety of diverse residential offerings. This includes both affordable and market housing, live-work and mixed-use units, and a range of densities. The highest density residential buildings are focused along 104th Ave. in anticipation of demand along the new LRT line. A few residential towers have been added (shown as striped circles) to mediate the existing skyline of the area, and further increase density in key locations. The existing, lower-density housing areas are preserved in places with tight-knit residential fabric. Gentle density intensification has taken place in these areas through the addition of laneway houses and other secondary units. Medium-density, multi-family housing has also been preserved throughout the site in areas where it is of higher quality and is offering an affordable option to residents. New affordable units are proposed to replace aging public housing, and other areas have new mixed developments that accommodate both market and social housing. Live-work units south of the mall are specifically intended for new immigrants.

	Entertainment/Nightlife		School
	Cultural & Recreational Hub		Religious Place

COMMERCIAL IDENTITY

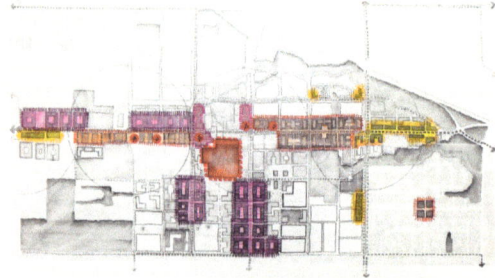

	Guildford Mall		High Density Mixed use
	Tower		Medium Density Mixed Use
	Small Businesses/Start-Ups		Corner Store

Two major factors drive the commercial sector design strategies. Guildford Mall and the proposed LRT are built upon in order to generate an urban hub.Urban life requires diverse activities and a sufficiently dense concentration of people. From this platform, mixed use opportunities are created with midrise buildings along the 104 Ave arterial. Also, the existing suburban, auto-oriented mall is activated along 104th Ave, with the addition of a new cultural hub. Activities are concentrated around the mall to provide job opportunities and support business start-ups. To support this, live-work typologies replace the mall's open parking along a section of 150th St and 152nd St. This area allows light industrial and commercial and retail uses for emerging small businesses. An entertainment zone with cafes and restaurants is suggested to the east side of the mall to keep the area alive during the night. Towers accommodate hotels, and small corner stores are added into residential areas. All of these diverse activities contribute to increasing vibrancy and economic opportunities for Guildford town centre.

PLACES AND CULTURE

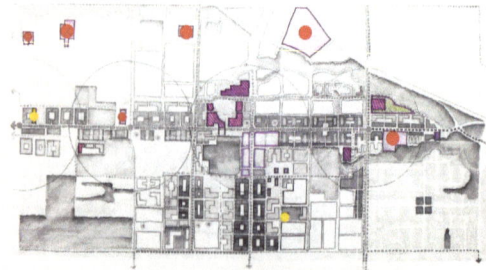

There are three strategies for enhancing the culture and identity of this place.

- Adding entertainment and recreational facilities. These provide richness and diversity of cultural activities and social interactions in both the day and night. The cultural hub and the mall co-function in order to catalyse everyday life.
- Connecting cultural hubs. The new cultural campus, and other entertainment nodes are connected by green spaces, pedestrian paths, and streets.
- Proposing multiculturally oriented programs. Schools, museums, churches, and other facilities are places that express different cultures and identities. All these facilities and places are very accessible by walking or taking transit. They contribute to the cultural mixture of Guildford as an arrival city.

ECOLOGY

Existing Park   Semi-Public Open Space
Preserved Green Space   Greenway
Public Open Space / Plaza   Bioswale

The ecological layer of the proposed urban design framework is based on the following strategies:

- Offering a green system, well-distributed in service of all;
- Improving accessibility to green/open spaces within 5 min (400m.) walking distance; and
- Revitalizing and preserving existing natural creek areas, watercourses and urban forests;
- Creating greenways to connect existing green areas and a series of proposed open spaces in order to shape a green loop.
- Developing urban agriculture and new farmers markets, distributed in the neighborhood.

Chloe Li   Amal Wasfi
Maryam Mahvash   Siyuan Zhao

As part of the ecological system, existing ditches offer opportunities to propose bioswales in the western end of the site.Taking advantage of semi-public open spaces of institutions such as schools and churches is another strategy to enhance the quality of Guildford's public realm.

MOVEMENT WATER AND ENERGY

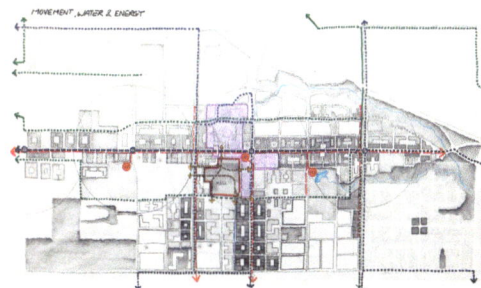

District Energy Plant   Bus Lane
District Energy Pipe   LRT
Proposed Green Line   Transit Stop
Existing Green Line

The movement network includes greenways, a new LRT line, and bus routes. The new LRT follows 104 th Ave, and has four accessible stops within the study area. The bus routes connect the rest of Guildford's neighbourhoods, and are especially concentrated in the core area: the mall and the recreation center. The movement of water relies on the natural Serpentine Creek system on the east side of Guildford. There are a large number of rainwater containers set in streets, green streets and plazas, parks, and natural riparian areas that collect rainwater and protect development from flooding. The district energy system in Guildford follows the main streets, and takes advantage of the waste heat coming from the mall site. The proposed system has three geo heat exchanges and provides efficient and sustainable energy to the area.

Prachi Doshi

Yashas Hegde

Nastaran E. Beigi

Saki Xueqi Wu

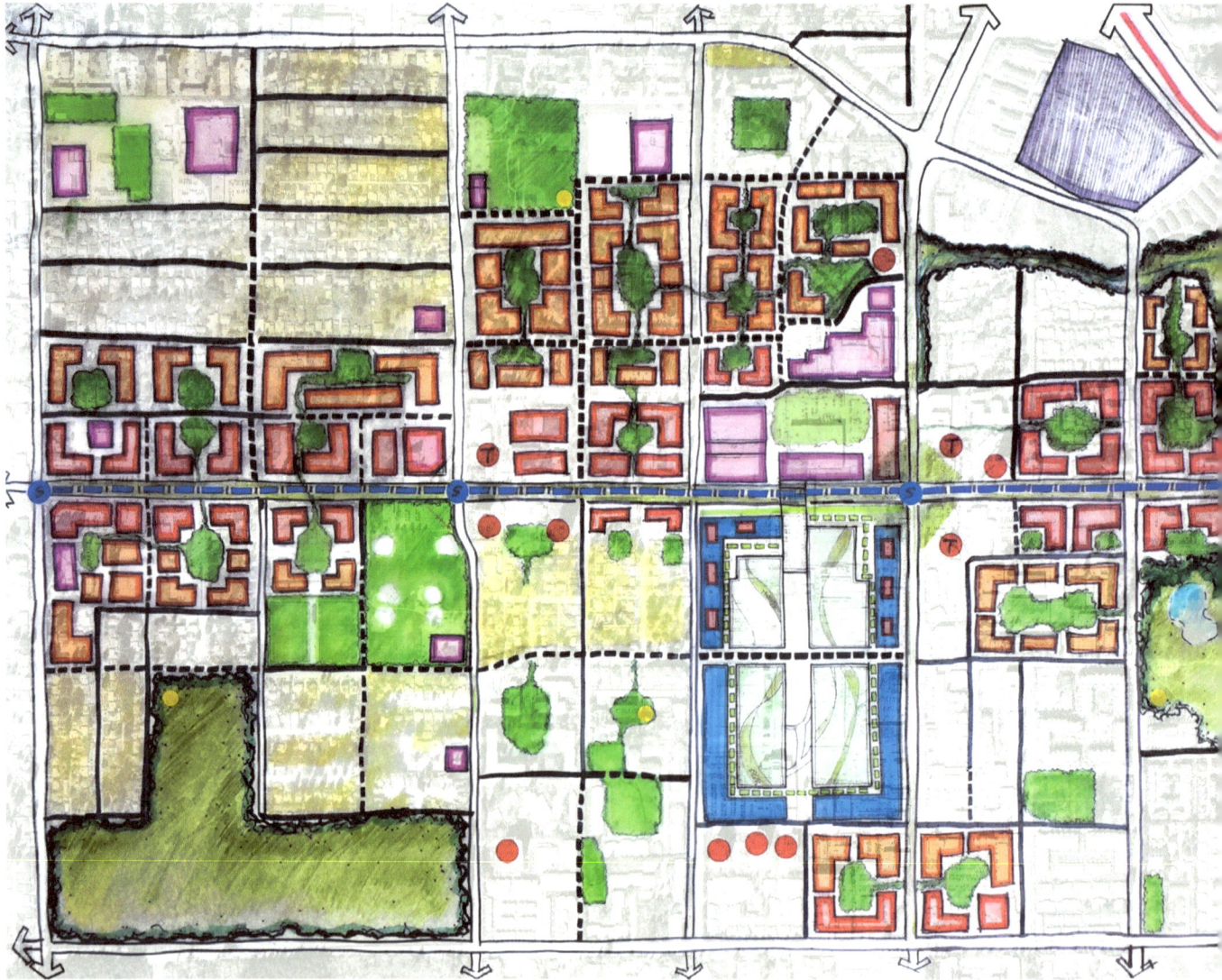

Hedieh Rashidi

Manali Yadav

Kate Wang

Lilian Zhang

Like many Surrey neighborhoods, Guildford has experienced a dramatic change over the past decade. It's an arrival place for new immigrants, young families and the aging population. In order to create an unique experience and a sense of place, the team was able to identify five main study areas as listed below:

a) The 104 corridor

b) The mall

c) Housing typologies

d) Green infrastructure

e) The auto-mall

Considering a vision for the next 40 years, this proposal aims to serve as an equitable and livable neighborhood for everyone. The new fabric focuses on the 104 Ave. with the proposal of light rail transit system.

WESTSIDE DESIGN

In the new proposal, mixed use buildings are proposed along the 104 Ave. towards the west side of the block. Additionally, mid-rise residential buildings are proposed behind the mixed use buildings. Courtyard style building typology is proposed in the scheme which emphasizes secured semi-private open space. This space tries to cut off the major arterial noise and acts as a safe enclosure for kids. The mall intends to reinvent itself by a phased proposal. The land economics will work efficiently by developing it over a period of time. The first phase proposal of the cultural center will be developed towards the north side of the mall. In the second phase, the theatre towards the south of mall will be eventually moved to the new location of cultural center. The third phase will place new mid-rise residential buildings over the podium along the periphery. The aim remains to open the mall by creating an axial corridor along the north-south axis of the mall. This will eventually connect to the cultural center. A few towers are located close to the existing towers.

EASTSIDE DESIGN

The north tip of the west side has an auto mall that will be transformed into a business incubator, while still retaining the auto shops at the ground floor. This will employ the local talent and the new immigrants and also add to the new economy. Interestingly, the urban forest and the park are preserved and the new development is planned along Serpentine Creek that seamlessly integrates within the wilderness. Mostly, mixed use and residential mid-rise buildings are proposed for the western four blocks. The overall scheme tries to retain most of the existing building while incremetally adding housing, commercial and cultural center for the Guildford region.

64

Prachi Doshi

Yashas Hegde

Nastaran E. Beigi

Saki Xueqi Wu

HOUSING STRATEGY

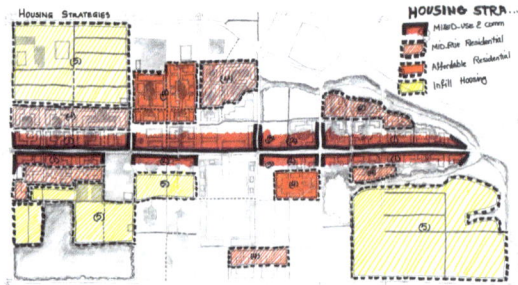

The proposed LRT system along 104 Ave. will potentially be a hotspot for new development and thus address the housing needs for the area. To accommodate the increasing demand for housing, the plan considers three main strategies. The first strategy proposes sensitive infill housing, which incrementally increases the density in neighborhoods (colored in yellow), while respecting the existing urban fabric. To be more specific, the gentle densification will be implemented in the under utilized areas (i.e. garages, empty backyards, etc.) by adding secondary suites, laneway housing and row houses. The second strategy encourages 4- to 6-storey midrise residential buildings (colored in orange and orange hatches). The proposed building typology is located within 5 minute walking distance from transit and commercial activities. Additionally, the major development is advised to occur on existing parking lots, vacant lands and old structures. Furthermore, some mid-rise residential buildings are allocated for affordable rental units on the existing rental fabric (colored in orange hatches). The third strategy recommends mixed use buildings along the 104 corridor. The majority of the proposed buildings are 6-storey mixed use buildings along with a few towers. The first and second floors are commercial and the rest are residential, providing services for community's daily needs.

COMMERCIAL IDENTITY

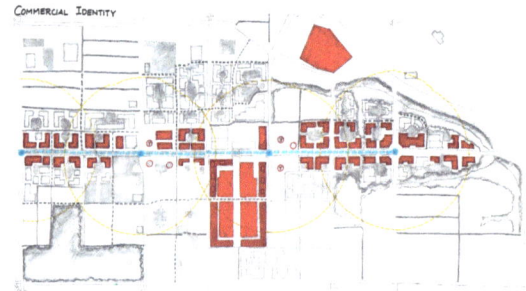

The strategy for commercial buildings was developed in service to the proposed LRT along 104 Ave. Guildford Mall serves as the heart of the neighbourhood. The commercial typologies form a crescendo at the mall which declines towards east and west. Proposed commercial buildings are complementary to the existing conditions. They are 4- to 10-storey courtyard style buildings along 104 Ave. to serve the residents within a 10 minute walking distance. Additional towers have been proposed near existing towers to create a symmetry in the streetscape. Higher density has been proposed at the LRT stations. Similar commercial buildings to north and south side facade of 104 Ave. creates uniformity and the mixed use buildings contribute towards retaining vibrancy. Apart from the commercial heart to the south of 104 Ave., we considered the auto-mall besides Highway 1 as a potential site for new immigrants to start businesses. Hence, the commercial identity represents a harmonious diversification along 104 Ave. with a concentrated density at the heart.

PLACES AND CULTURE

The cultural and recreational hub located towards the northern side of the mall acts as a prime location for cultural activities. This hub serves as a place to learn, relax and play for the residents. The focus of this proposal encourages locating activity nodes within 5 minute walking radius of these eight blocks. These activity nodes are comprised of community garden, farmers market and play areas for kids. They are located within the green areas and are easily connected through the greenways. The existing schools, churches and religious centers effectively serve the current population. Connecting them through the green corridor encourages residents of all ages and abilities to use it. Our earlier research suggested we allocate more neighbourhood houses across these eight blocks. Additionally, daycare will be located within the housing complex. Hence, residents do not have to rely on their private automobiles to use such facilityies in the neighbouring districts. Public art is located in a prime location and also within the parks and playgrounds. This conveys the historic and cultural pride and educates the new immigrants about their new adopted place. The plan finally focuses on respecting the diverse cultural and social values brought by the new immigrants and seamlessly integrating them with the existing fabric.

Hedieh Rashidi   Kate Wang

Manali Yadav   Lilian Zhang

ECOLOGY SYSTEM

In our proposed plan, urban forests, green urban parks, river corridors, green courtyards and community gardens form a hierarchy of urban green spaces. These green spaces have been integrated and connected through a green corridor network.

First, the plan preserves natural resources like urban forests and salmon habitats. Second, the environmental connectivity of urban green spaces is promoted by green corridors. The main objective of the green corridor network is to connect the neighbourhood green spaces and promote safe pedestrian and bicycle routes. Also, the network is linked to a sustainable street pattern to encourage people to travel to work, school or to local services by foot or bicycle. Furthermore, the plan proposes several community gardens which are close to the parks and forest area. These community gardens are located within the existing parks and supply the surrounding neighbourhoods. This new type of urban agriculture will help contribute to food security for this region.

MOVEMENT WATER AND ENERGY

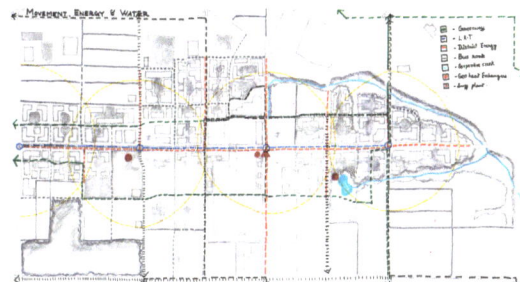

The above diagram illustrates the conceptual ideology for water, energy and movement. The light rail transit investment on 104 Ave. has significant impact on the use of land parcels along the 104 Ave. The basic need to shift from the current typologies to compact urban typologies is an inevitable urban design strategy. Consequently, this strategy ensures a more energy efficient use of the district energy line along the 104 Ave. It is economically not viable to connect district energy to buildings that are old, hence the district energy lines do not extend to zones where development of land is minimal. Geo-heat exchangers ensure a positive energy dissipation which results in a smarter energy grid. This strategy aids in creating an energy loop around the Guildford mall which shall be dissipated to the surrounding parcels of land that uses significantly less energy. The district energy system shall get connected to the Sewer Heat Recovery Plant at Port Kells.

66

Chapter 5: Urban Framework

Building on the shared site visit and intuitive first impressions, and informed by Surrey's policy and best practice intentions, the studio continued to iterate urban framework investigations into a single, shared vision for the future of Guildford. The studio experience, while compressed, made time for the important rigor of design iteration to simulate the effective methodologies that are required when working with complex stakeholder groups with conflicting values and intentions. Once a synthesis framework was identified, the studio organized into teams with each producing respective layers of the vision. These technical layers were developed with each team mindful of the overall shared framework thus allowing for efficient, obvious and coordinated integration. The studio established shared graphic conventions that would allow the final urban framework for Guildford to be produced quickly and professionally.

⑤ PROPOSED URBAN FRAMEWORK

guildford :: YES

Early investigations on the Guildford precinct

During the course of 2014 Fall term, the first cohort of the Masters of Urban Design Program at the University of British Columbia investigated the regional context and growth strategies for the City of Surrey, British Columbia at a city scale. The Surrey officials requested an urban design framework for the Guildford region with a proposed light rail transit system. The 2015 MUD Winter studio will further investigate Guildford at neighbourhood scale based on the research of the last studio. The size of the Guildford study area is two miles by one mile. This precinct is characterized by "varying, complex, urban structures, related built form/typologies, a large scale economically viable mall that has recently enjoyed substantive re-investment, an active small business community, distinguished open space and natural landscape/water systems, amenities and certain housing affordability allowing entry into the Canadian west coast market. The Guildford Precinct is recognized as a "market entry portal" that distinguishes it as an 'Arrival City'." (Hein 2015)

The urban design framework of this project is achieved through team collaboration. First, eight small groups performed site investigations, documentations, and analysis for different blocks of this precinct. Thereafter, the studio was divided into two groups and started with design iterations based on the shared existing conditions. Students met with Surrey officials, academics and economic advisors and gained expert advice on myriad topics. The final proposal reflects how innovations and interdisciplinary rationals work together. The overall framework is followed by sixteen individual investigations that "further reveals strategic implementation potential" (Hein 2015) on different pieces of the precinct.

Figure 1: The Guildford precinct

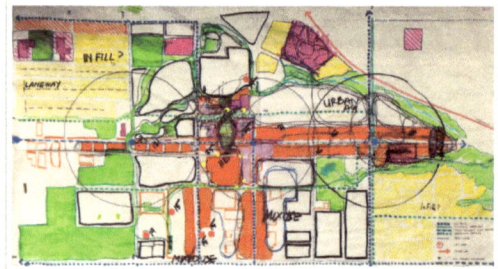

Figure 2: Analysis and Investigation: Team Scot

Figure 4: Further design test: Team Scot

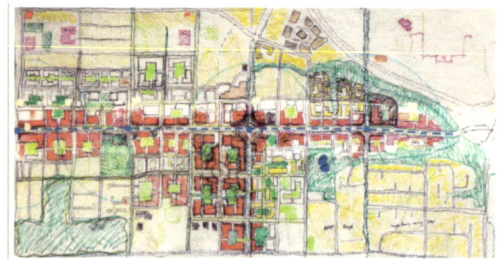

Figure 3: Analysis and Investigation: Team Don

Chen Fan

A quick view of the urban design framework

The urban design framework proposes a transit-oriented development, which will promote the gateways on both sides of the precinct as "welcome mats" and the mall area as a "central urban heart". The proposal integrates green space, natural assets, and the food system into a "green loop", which advocates an ecologically-friendly and healthier lifestyle. Moreover, taking the metaphor of "quilting", the plan reshapes and renews some existing development patterns. A lot of new housing, commercial, service and cultural opportunities are added, which will encourage more multi-cultural, democratic and vibrant everyday life. The following sections will cover all the proposals in detail.

Figure 5: Design idea: Team Don

Figure 6: Final concept

Figure 7: Final master plan

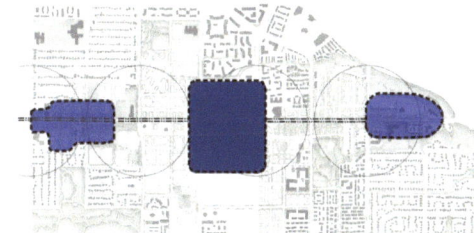
Figure 8: Welcome mat

Figure 9: Cultural pearls

Figure 10: Green loop

72

Figure 11: Contextual quilt

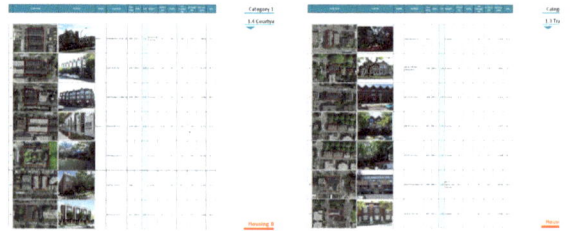
Figure 12: Housing opportunity

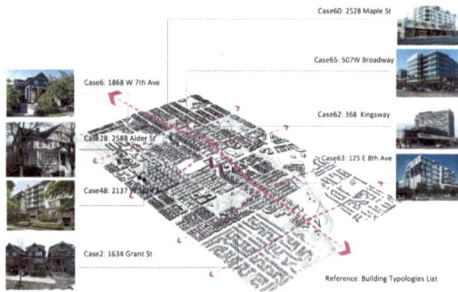
Figure 13: The new image and scale of the city

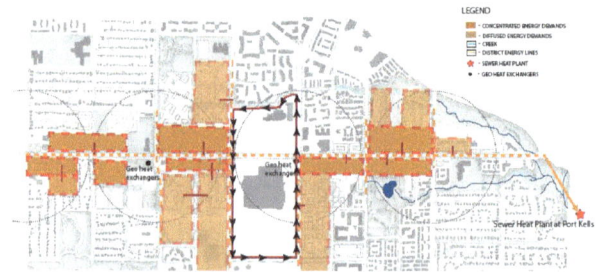
Figure 14: Energy system

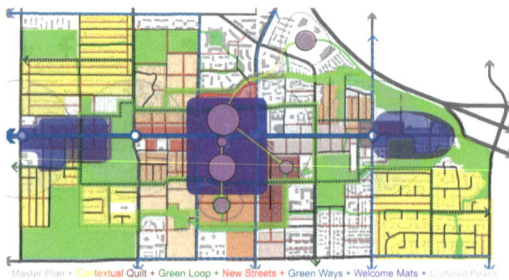
Figure 15: Integration of all the layers

Reference
MUD Studio Overview, S.Hein, Winter Term, 2015

73

Avishan Aghazadeh

1- Our Graphic Identity for Guildford

We have created this identity to distinguish our design intentions and also to reflect our humanist values that embrace the important role that Guildford will continue to serve as an Arrival City.

2- Our fundamental urban design drivers:

* Nature-Hoods
* Transit-Oriented Culture
* Contextual Quilting

guildford :: *YES*

* *Humility and Humanity*
  As represented by lower case
* *Diverse, Vibrant and Playful*
  As represented by varying fonts and colors
* *A Place of Design*
  As represented by our centralized cross axis, shown with double colon
* *New Opportunity energized by the Promise of Transit*
  As represented by the dynamic *YES*

3- Our Values:

* Arrival City:
  We have focused on various aspects of the Arrival City such as demographic diversity, housing and economic opportunities.
* Scale and context:
  We have tried to acheive the density with building typologies that respect the existing context.
* Incremental development:
  We have proposed gradual infill through different phases of time, both to respect the existing fabric and also to preserve the character of the neighbourhoods.

Over the past 12 weeks, we have worked collaboratively as if we were a professional consultancy office to produce what is equivalent to one year's worth of hard work.

The first thing we did was we divided Guildford area into eight blocks.

We visited and investigated each block in groups of two.

We sketched and played a musical composition reflective of the existing north and south side streetscapes of 104 Ave. as a way to sharpen our contextual observation and appreciation.

Haneen Abdul Samad

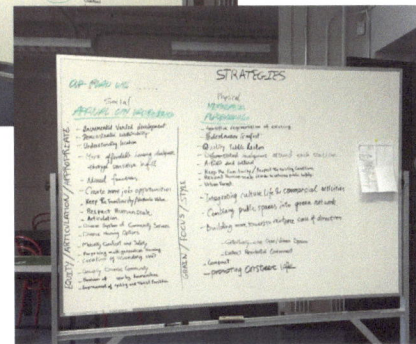

Meetings with planners, designers, policy makers and transit experts lead to final decision along with several design charrettes.

To do justice to our work we tried to find the most appealing words that described our design and intentions.

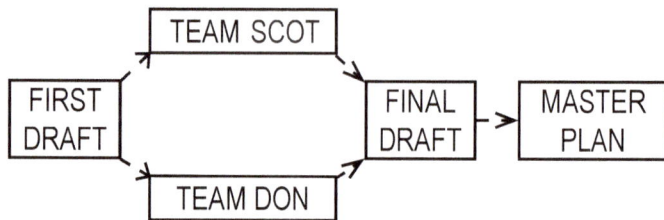

PROCESS

Guildford's framework is discovered through iteration towards synthesis. First, the studio worked together and integrated a goal for the project. Then, the 16 students were split into two teams and each team created different design results. Combining and weighing the pros and cons of all the previous work, the class came to a final master plan.

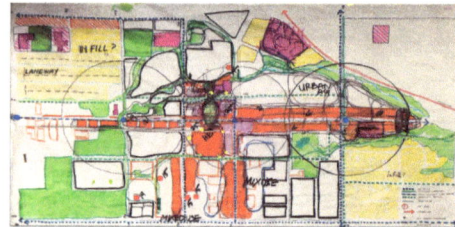

PHASE 2 - TEAM SCOT

Team Scot's idea was to keep the mall, and enhance the vibrancy of the area with some entertainment features.

HOUSING STRATEGY   COMMERCIAL IDENTITY

PHASE 1 - FIRST DRAFT

The process was intiated by deciding what to preserve and enhance. The goal for the Guildford region is to add vibrancy to the 104 Ave. and focus on the commercial hub i.e. the mall.

PHASE 2 - TEAM DON

Team Don's strategy was to incrementally replace the mall with some smaller scale comercial features and open it up.

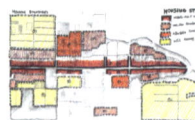 

HOUSING STRATEGY   COMMERCIAL IDENTITY

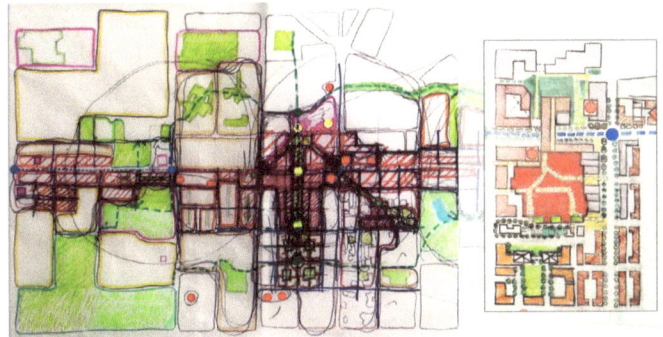

PHASE 4 - FINAL DRAFT

Based on Team Scot and Team Don's work, the final draft opened up the N-S axis by creating a hub and designing more connections with the surrounding environment such as the cultural center and green space.

S AND CULTURE · MOVEMENT, WATER AND ENERGY · ECOLOGY

PHASE 5 - FINAL MASTER PLAN

In the final master plan, the team broke down the details of the draft and refined our ideas and strategies for the Guilford region.

S AND CULTURE · MOVEMENT, WATER AND ENERGY · ECOLOGY

A Humanized Transit Corridor
Welcome Mats an the Heart

104 Ave. is the main transit corridor in the neighbourhood of Guildford. A light rail transit is proposed on 104 Ave. until 156 St. The strategic location of the vibrant mall on the transit corridor enhances its ability to be the heart of the neighbourhood. The two welcome mats act as gateways for the region. These mats serve as an opportunistic feature, also focusing on the new immigrants.

■ The Welcome Mats

■ The Guildford mall

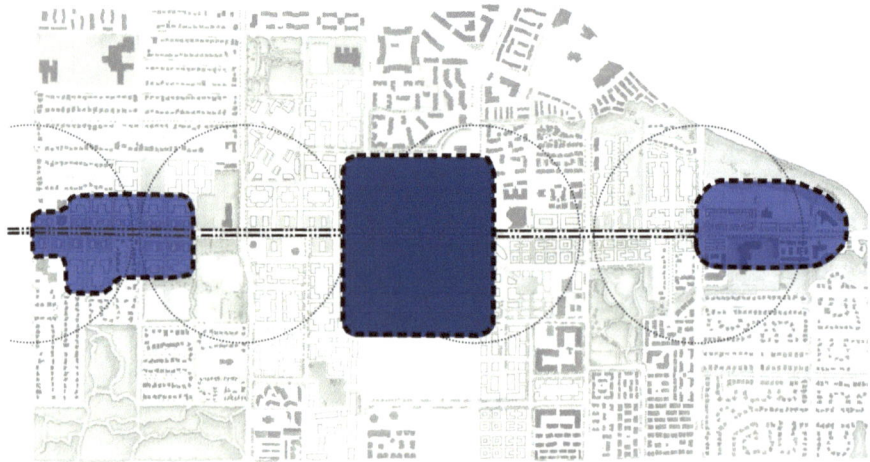

Fig. 1: 104 Avenue, with Guildford mall at the centre as the heart of the neighbourhood and the east and west welcome mats

The Green Loop

Green areas are a treasure to the City of Surrey. Almost 34% of the land coverage is comprised of the ALR / green spaces. Green Timbers Urban forest to the southwest and the Guildford Brook to the northeast of the study area, leave a possibility to connect the eight blocks. The basic idea here is to encourage pedestrians to use the natural spaces which would be complemented by a significant reduction in automobile industry. The addition of green spaces considering the existing topography serves the environmental concerns as well.

■ The green loop

Fig. 2: The existing and the proposed green spaces, interconnected with the greenways, parks and the existing forest areas

Prachi Doshi

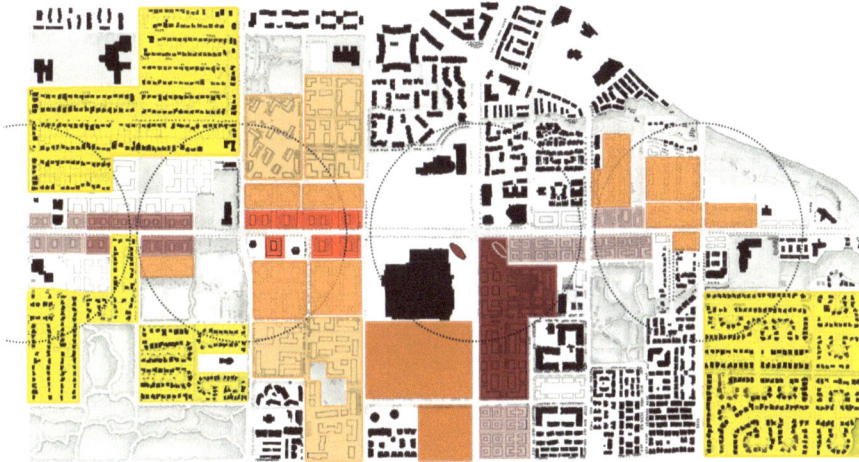

Fig. 3: The contextual quilt replicating mosaics of land-use

The Contextual Quilt

It considers aspects such as Guildford's existing development, ethnicity of the neighbourhood, energy usage, environmental issues, cultural necessities, commercial opportunities and future possibilities.

- Gentle intensification
- Multi Family Residential Refresh
- New Multi Family Residential 1
- New Multi Family Residential 2
- Mixed Use 1
- Mixed Use 2
- Mixed Use 3
- Commercial Refresh

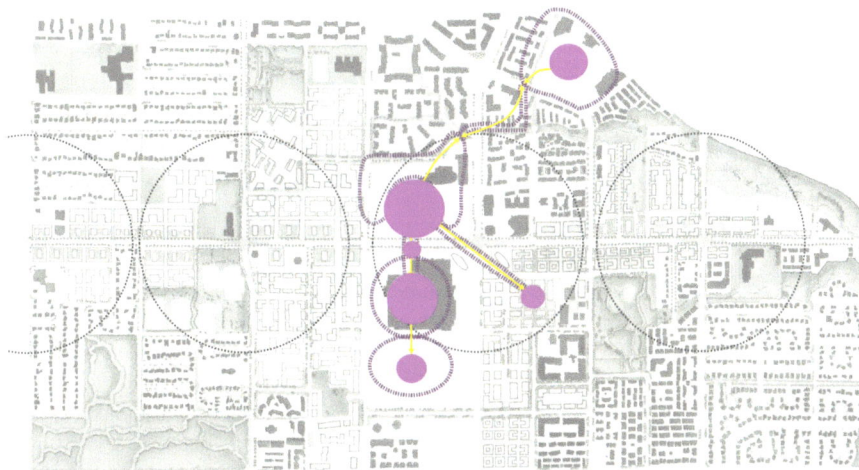

Fig. 4: The existing and the proposed cultured pearls which are interconnected to make ithem more accessible

The Cultured Pearls

Surrey as a City amalgamates various ethnic groups. People from various cultures are known to form communities and settle in Surrey. For it to be an Arrival City, various cultural elements need to satiate the demands of these communities. Guildford exhibits such few precious features. Apart from that, new features have been proposed including the area to the east of the mall, known as 'The Guild', which provides opportunities for various activities. These cultural elements range from a physical to an experiential space.

- Cultural Pearls

Manali Yadav

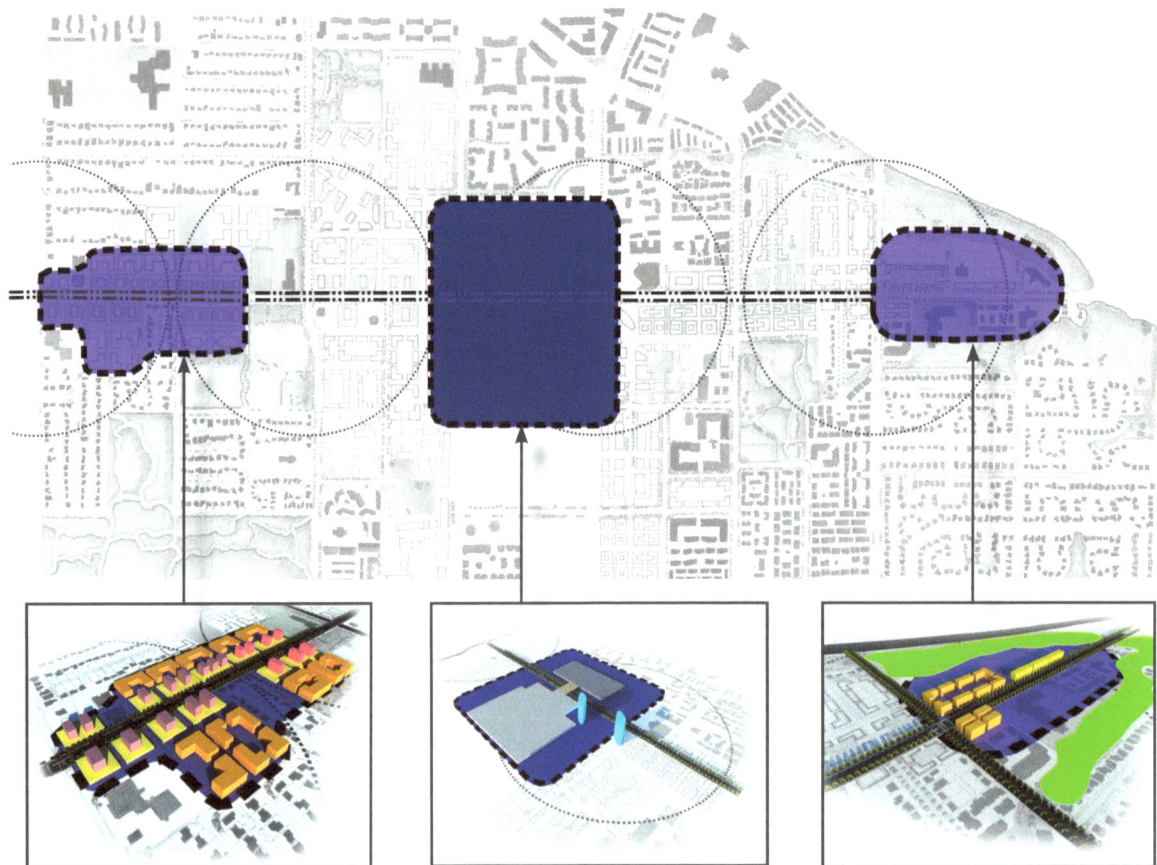

Fig.1: West Side welcome mat

Fig.2: Heart of Guildford - the mall

Fig.3: East Side welcome mat

The vision of this project aims to create a more humanized, walkable and visually-engaging corridor, and the arrival of light rail transit in the region will facilitate this transition. The highlighted region that forms the two ends of the quilt are the welcome mats. The mats serve as gateways to the area, and new developments within these welcome mats give cogency to the corridor's visual signature through human scale frontages and a new rhythm of mid-rise residential buildings (Fig.1). The design is occasionally interrupted by natural features such as urban forest, streams and parks (Fig. 3). Furthermore, the mats deliver a more local and distinctive community experience for residents and visitors. Similarly, the mall situated at the heart of the quilt acts as a cultural and commercial center for Guildford (Fig. 2). In summary, the design creates a dynamic pedestrian experience by strengthening the social, cultural and business realms within Guildford.

Simone Levy

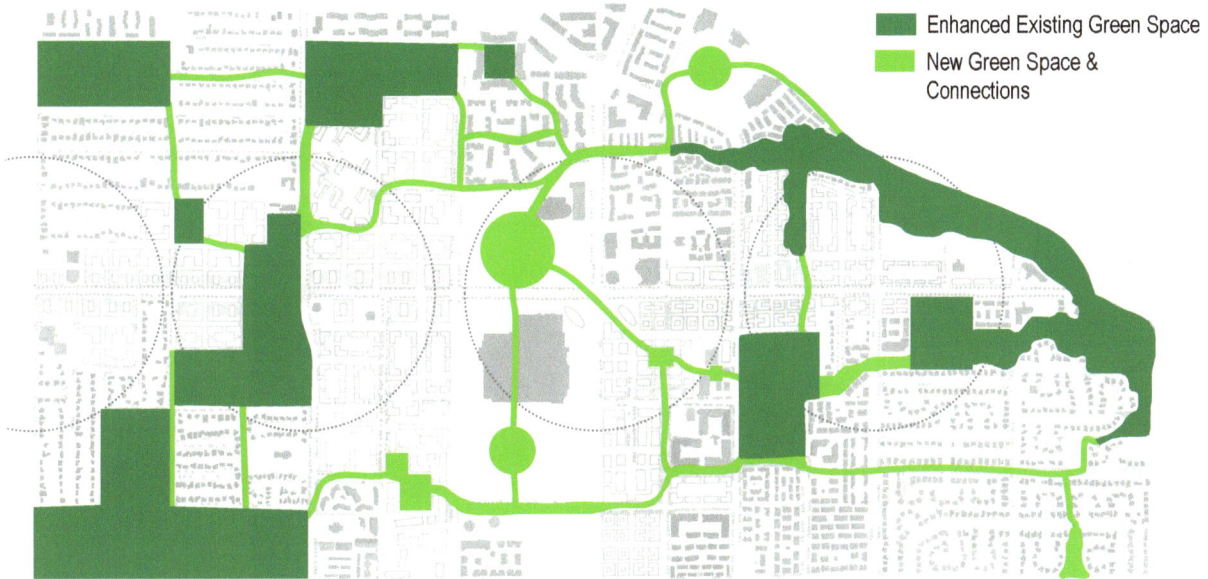

Enhanced Existing Green Space

New Green Space & Connections

Guildford's proposed Green Loop offers a special opportunity to improve upon the area's existing natural resources and connect them in order to make them a prominent part of the neighbourhood's new urban framework.

Existing green spaces in the study area include parks, school fields and Serpentine Creek. In addition to enhancing the health and biodiversity value of these areas, the plan adds new public spaces through the centre of the site in order to connect the northern and southern sysems. This also adds much needed open space for the new residents in the higher density developments across the study area.

The entire Green Loop acts as a place for recreational opportunities, community self-expression, continuous habitat and the movement and production of local food. The connected system is meant to become a unique asset for Guildford, and a destination for residents of the entire region along the new 104 Ave. LRT line. The Green Loop also serves as a means for connecting the individual neighbourhoods within Guildford to each other, and to the central commercial and cultural zone of the area.

82

The Contextual Quilt

This map shows proposed residential and commercial buildings for the Guildford neighbourhood. The contextual quilt is comprised of a series of patches that indicate different uses. The patches are coloured in different tones, each representing a particular strategy for that area.

Black: Areas to be preserved (status quo).

Yellow: Gentle intensification areas to undergo low-level intensification through an infill strategy, featuring either lane ways or culs-de-sac.

Very light orange: Multi-family residential refresh: multi-family residential housing to be built on existing footprints or to conform to new construction.

Medium orange: Multi-family residential No.1: multi-family residential housing to be built on non-arterial roads.

Dark orange: New multi-family residential No. 2: areas with tremendous potential for new development to introduce new units.

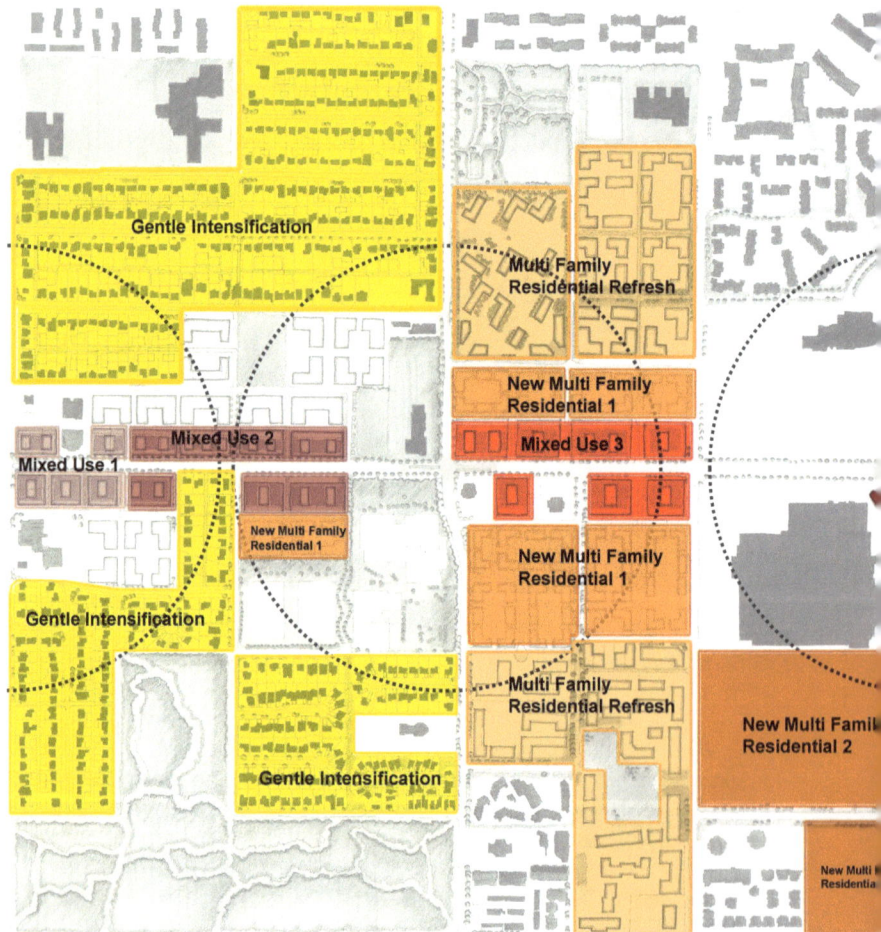

Gentle Intensification

Multi Family Residential Refresh

New Multi Family Residential 1

Mixed Use 2

Mixed Use 1

Mixed Use 3

New Multi Family Residential 1

New Multi Family Residential 1

Gentle Intensification

Multi Family Residential Refresh

New Multi Family Residential 2

Gentle Intensification

New Multi Family Residential

Nastaran E. Beigi

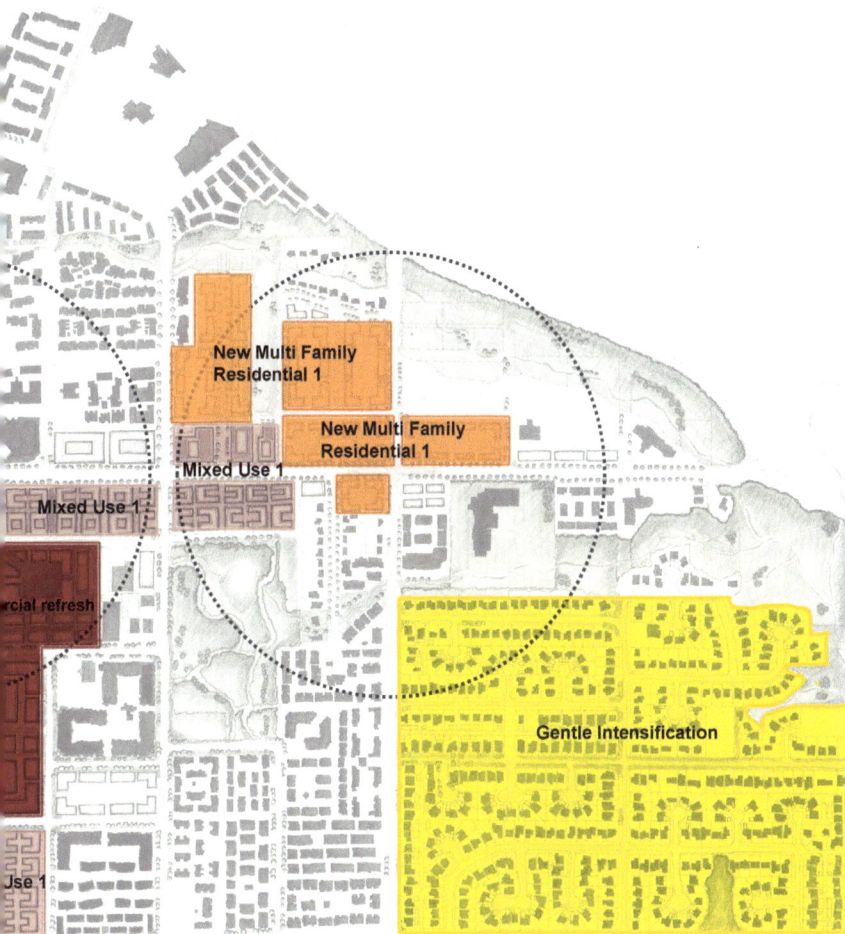

New Multi Family Residential 1

New Multi Family Residential 1

Mixed Use 1

Mixed Use 1

cial refresh

Gentle Intensification

Jse 1

Light red: Mixed use No. 1: mid-rise, six-storey, podium-style building, with the bottom two floors allocated for commercial use and the remainder for residential use.

Medium red: Mixed-use No. 2: mid-rise, eight-storey, podium-style building with the bottom two floors allocated for commercial use and the remainder for residential use.

Dark red: Mixed-use No. 3: mid-rise, 12-storey, podium-style building with the two bottom floors allocated for commercial use and the remainder for residential use.

The height of these buildings establishes a hierarchy. Interestingly, their height increases with distance from the edge of the site.

Darkest red: Commercial refresh, with no residential or other uses permitted.

Reference:
Resource: UBC Geography Information Center

84

Fig 1: Site Location

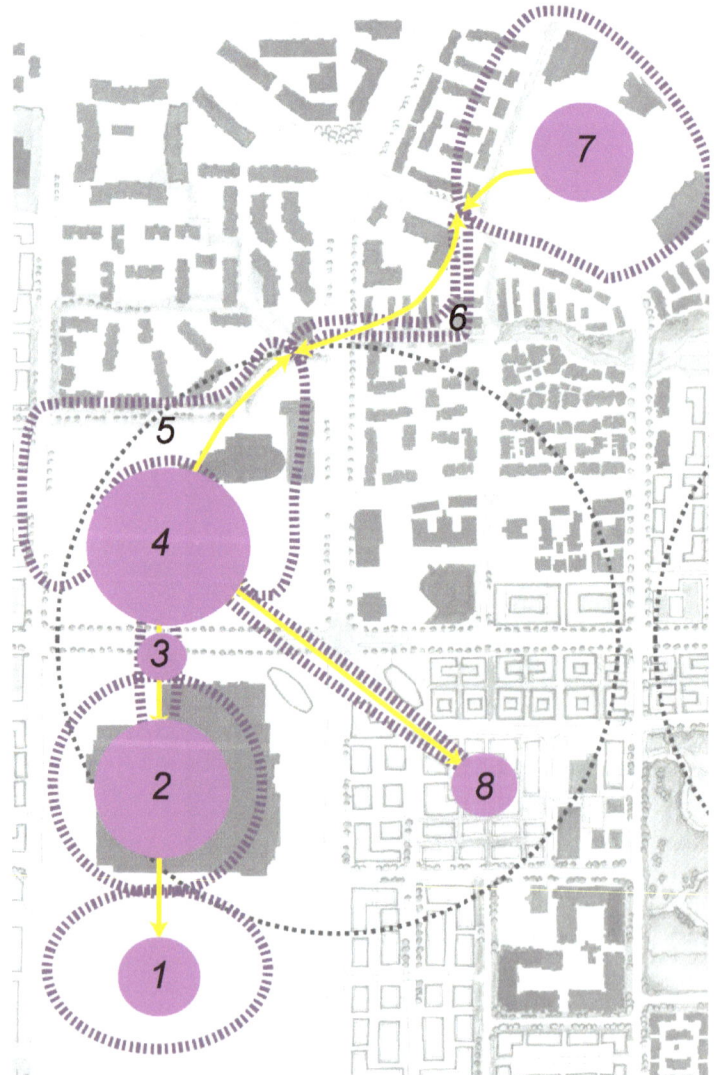

Fig 3: The Mixed neighborhood precedents

This diagram shows cultural pearls on the site, with each pearl representing distinct cultural characteristics. (Fig 2)

1. The first pearl is the mixed neighbourhood, which provides 30% affordable rental units and shared studio spaces. This means the creation of market housing on the site of former parking lot (Fig 3).

Fig 4: The Guilford Mall

2. The next pearl is the mall. It is celebrated as the existing cultural hub of the Guilford area until it is ready to utilize its full capacity for future renovation (Fig 4).

Fig 5: Veccio Bridge and New York Highline- The Transit Bridge precedents

3. The transit bridge is the next cultural pearl that has the potential to become a unique architectural feature of the area. The most relevant visual reference could be Veccio Bridge and New York Highline (Fig 5).

Fig 2: Culture Pearls and The Guild

Hedieh Rashidi

Fig 6: Pritzker Park, Chicago-The Cultural and framing the commons precedent

Fig 7: The Art Walk precedent

Fig 8: The Cultural Campus precedent

Fig 9: The Guild Precedent

4 & 5. Next is the cultural commons. It is located on the north side of 104 Ave. Created as a new central open space for the region, it is big enough to hold concerts, ceremonies, and citywide festivals. The next pearl is complementary to the cultural commons called Framing the Commons. It is a new cultural infrastructure including relocated theatre, concert hall and public art. The most relevant visual reference is the Pritzker Park in Chicago (Fig 6).

6. The Art Walk represents another pearl, as a pedestrian connection between the commons and the cultural campus (Fig 7).

7. The cultural campus is the site of the former auto mall, which has the potential to be re-imagined as a creative hub. For instance, institution, film industry, production facilities, etc. (Fig 8).

8. The last pearl is The Guild. A Grandville Island-like neighborhood that signifies another cultural destination. This incremental open space will accommodate local activities including gatherings and festivals (Fig 9).

Contextual Housing Opportunity

Reviewing Vancouver Housing Catalog's Best Practice Case Studies, Guildford's housing opportunity is categorized into nine categories as follows:

Status Quo: Areas designated relatively young age of development, and/or challenges of unwinding existing strata corporations, and related market rental capacities that should be preserved. No new units are included in the speculative housing capacity total.

Laneway Gentle Intensification 1: It introduces laneway single family dwellings, laneway duplexes, laneway triplexes and laneway multiple dwellings. The density ranges from .6 to 1 FSR/two-storey height with 335 units anticipated at an average UPA of 30.

Laneway Gentle Intensification 2: It introduces culs-de-sac oriented in addition to live-work on fronting driveway footprints contiguous with internalized garage volumes. The density ranges from .6 to .75 FSR/two-storey height with 160 units anticipated at an average UPA of 10.

Multi-Family Residential Refresh: It intensifies, without displacement, existing older market/rental housing stock by incrementally upgrading with more intensive wood frame format on existing footprints with additions. The density ranges from 1.5 to 2.5 FSR/four to six-storey height with 2000 units anticipated at an average UPA of 60.

New Multi Family Residential 1: It replaces under-developed areas within 5 minutes walkability to new transit investment. Proposed typologies are reflective of local scale/frontage and are intended to strengthen community and social exchange. The density ranges from 1.5 to 2.5 FSR/four to six-storey height with 300 units anticipated at an average UPA of 75.

New Multi Family Residential 2: It is a distinct, residentially intensive precinct with predominantly market housing (towards value creations and mall transformation of existing parking lots/relocation of the Cineplex). The density ranges from 2.0 - 3.0 FSR/four to eight-storey height with 885 units anticipated at an average UPA of 60.

13. Arbutus Walk
14. 1820 Bayswater Street
15. 507 West Broadway
16. 125 East 8th Avenue

Mixed Use 1: It replaces under-developed and arterial fronting areas within 5 minutes walking to new transit investment. Proposed typologies are reflective of local scale/frontage and are intended to strengthen community and social exchange while minimizing livability impacts of transit corridor adjacency. Residential mid-rise slab buildings atop a commercial podium are oriented away from the arterial noise while also maximizing southerly natural light to penetrate the 104 Ave. corridor and residential blocks further north. The density ranges from 2.0 to 2.5 FAR/two to six-storey height with 1459 units anticipated at an average UPA of 90.

Case Studies

1. 1868 West 7th Avenue
2. 320 West 15th Avenue
3. 1634 Grant Street
4. 560 Hawks Avenue
5. 2632 Hemlock Street
6. 1347 West 7th Avenue

7. 2588 Adler Street (@ West 10th)
8. 2052 Cypress Street
9. 700 West 12th Avenue
10. 2250 West 10th Avenue
11. 2036 West 10th Avenue
12. 2137 West 10th Avenue

17. 2770 Sophia / 368 Kingsway
18. 2528 Maple Street
19. 2515 Ontario Street
20. 320 west 15th Avenue
21. 1634 Grant Street

Reference: Housing Best Practice Case Studies, City of Vancouver.

Maryam Mahvash

Status Quo

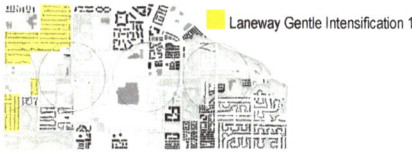
Laneway Gentle Intensification 1

Cul-de-sac Gentle Intensification 2

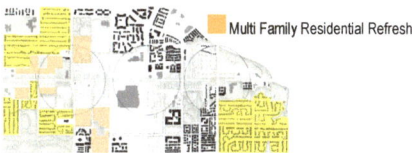
Multi Family Residential Refresh

New Multi Family Residential 1

New Multi Family Residential 2

Mixed Use 1

Mixed Use 2

Mixed Use 2: It replaces under-developed, arterial fronting areas within 5 minutes walking distance to new transit line. Proposed typologies are reflective of local scale/frontage and are intended to strengthen community and social exchange while minimizing livability impacts of transit corridor adjacency. Residential mid-rise slab buildings atop a commercial podium and are oriented away from the arterial noise while also maximizing southerly natural light to penetrate the 104 Ave. corridor and to the residential blocks further north. The density range from 2.5 to 3.0 FSR/two to six-storey height with 900 units anticipated at an average UPA of 100.

Mixed Use 3: It replaces under-developed, arterial fronting areas within 5 minutes walking distance to new transit line. Proposed typologies are reflective of local scale/frontage and are intended to strengthen community and social exchange while minimizing livability impacts of transit corridor adjacency. Residential mid-rise slab buildings atop a commercial podium are oriented away from the arterial noise while also maximizing southerly natural light to penetrate the 104 Ave. corridor and residential blocks further north. The density ranges from 3.0 to 3.5 FSR/two to nine-storey height with 3000 units anticipated at an average UPA of 110 (Relavant cases: 15, 16, 17, 18 & 19).

Status Quo | Laneway Gentle Intensification 1 | Cul de Sac Gentle Intensification 2 | Multi Family Residential Refresh

New Multi Family Residential 1 | New Multi Family Residential 2 | Mixed Use 1 | Mixed Use 2 | Mixed Use 3

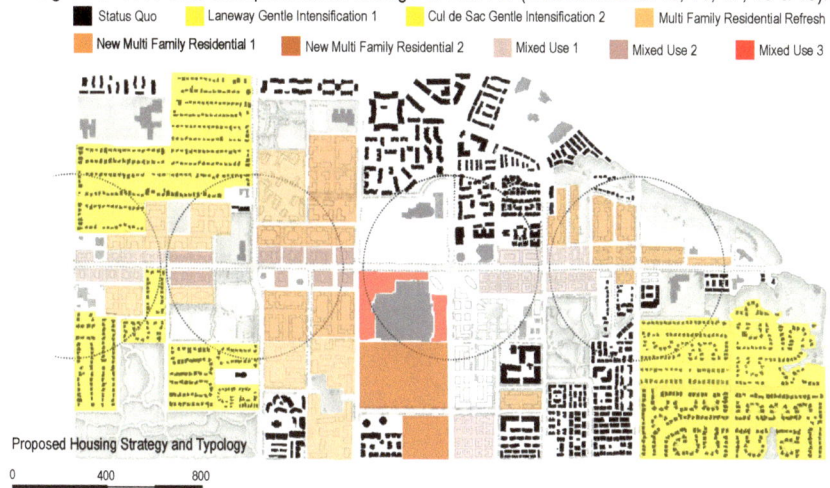
Proposed Housing Strategy and Typology

0   400   800

Point towers were considered as a strategy to create economic value that could be attributed to local amenity in the early design process. However, they will not be as successful as in other contexts and further, they symbolically compromise our intentions for Guildford to remain as an Arrival City. Point towers may motivate higher land costs given expectations by land owners should they be introduced. Guildford must remain relatively affordable as an Arrival City. Hence there are no residential point towers in this proposed urban framework. Notwithstanding, our total anticipated new housing units is approximately 10000 units.

FSR	Height (Storey)	UPA	Units
3-3.5	2-9	110	3000
2.5-3	2-6	100	900
2-2.5	2-6	90	1459
2-3	4-8	60	885
1.5-2.5	4-6	75	1300
1.5-2.5	4-6	60	2000
0.6-0.75	2	10	160
0.6-1	2	30	335
Total units			10000

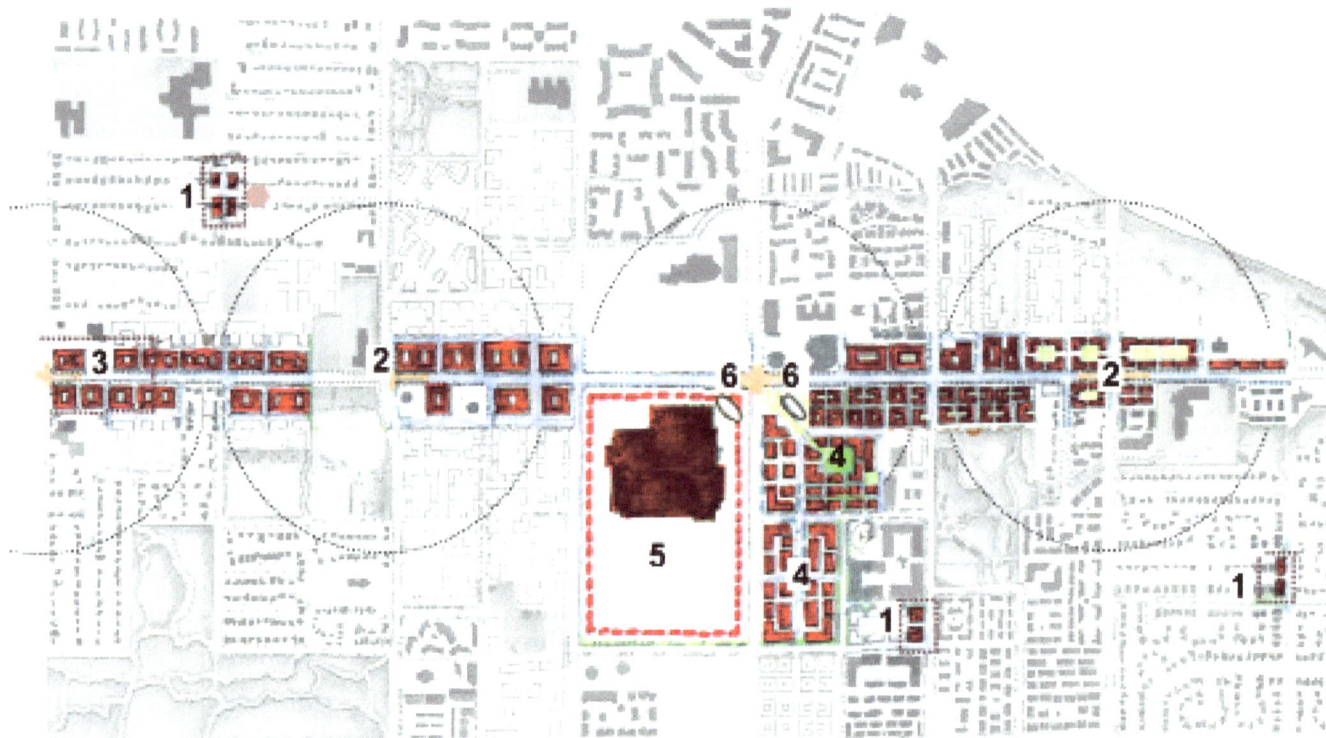

Map showing the different commercial strategies

1- Mom And Pop Down:
Small retail on the Neighborhood scale

2- Transit Oriented Retail-General:
Mixed use buildings across the arterial

3- Auto Focused Retail.
Moving "Car R Us" to the main arte

"The Guild" Area
iness "incubator" for new immigrants/start-ups.

5- "The Big Pearl" (Guildford Mall)
Considered re-configuration to create high street experience.

6- "Twin" office/hotel tower
The towers will visually announce the "heart" of Guildford

Amal Wasfi

Our proposed commercial strategy on varying scales are as follows:

1- Mom and Pop Down on the Corner - Small scale, local-serving retail that includes fresh produce, milk, coffee and a pub, all within 5 minutes walk.

2- Transit Oriented Retail - General stores at varying frontages and typically associated with mixed use residential development opportunities.

3- Transit Oriented Retail - Special 1 "Cars R Us" associated with existing car lots and to create a branded retail experience with long, yet vibrant frontages and to facilitate decanting of the northerly, suburban style Auto Mall in favor of a special "Cultural Campus".

4- Transit Oriented Retail - Special 2 (Commercial Refresh) AKA "The Guild" that is intended to become a special, small scale, business "incubator" precinct for new immigrants/start-ups, that could accommodate incremental increase from idea to pilot to production scale, a visual place of making, light (non-noxious) industrial and generally unique, gritty and a distinctive community of innovators. The precinct is anchored by a tall tower/hotel that emphasizes the connectivity to the commercial patronage of the mall and the new Cultural Precinct north of the mall.

5- Transit Oriented Retail - Special 3 "The Big Pearl" (Guildford Mall) that is recognized as the commercial, and perhaps cultural, heart of the community and is positioned as the largest of a string of commercial/residential/public open space and amenity/cultural string of pearls about the north-south axis of the existing bridge. The mall is viewed as an important asset to the local, and transit oriented larger community, and could have many futures including shifting existing floor plates/tenancy to a more urban fronting relationship with the 104 Ave. Transit Corridor, or promote visual engagement of the mall's outer edges, or a re-configuration to a pronounced axial arrangement, with possibly exterior public open space or high street experience contiguous with the existing bridge. Recent substantive investment is recognized. The mall's future is economically strategic towards the introduction of improved bridge infrastructure/visual identity and a strengthened cultural and open space precinct further north.

6- "Twin" Office/Hotel Towers - To assist with the above listed economic strategy, we are proposing the introduction of a single, large floor plate office and/or hotel tower of a height up to 300' to be located on the northeast corner of the site. The plan anticipates a floor plate shape orientated to minimize shading impacts on the public realm. Further, this location is preferred given the "pairing" with a "twin" office/hotel tower immediately east as both will visually announce the "heart" of Guildford in the larger Surrey region. This would occur, along with the anticipated new amenities, and immediately adjacent special cultural/commercial precincts.

1.Educational&Cultural Route

Legend

- Cultural Hub
- The Art Walk
- Recreation Institution
- Educational Institution
- Religious Institution

1. Educational & Cultural Route: Guildford as a "place of culture", the strategy is to build on recent cultural and recreational investment on the north side. There is potential to decant the auto mall by replacing it with a vibrant Cultural Campus.

2.Religous&Cultural Route

Legend

- Cultural Hub
- The Art Walk
- Recreation Institution
- Educational Institution
- Religious Institution

2. Religious & Cultural Route: Guildford accommodates a large number of diverse immgrants, this route represents the cultural diversity.

Xinyun Li

3.First Nation Cultural Route: The east side of Guildford, the existing natural forest and green space, more local landscape and culture represents the local culture though sculptures and art installations.

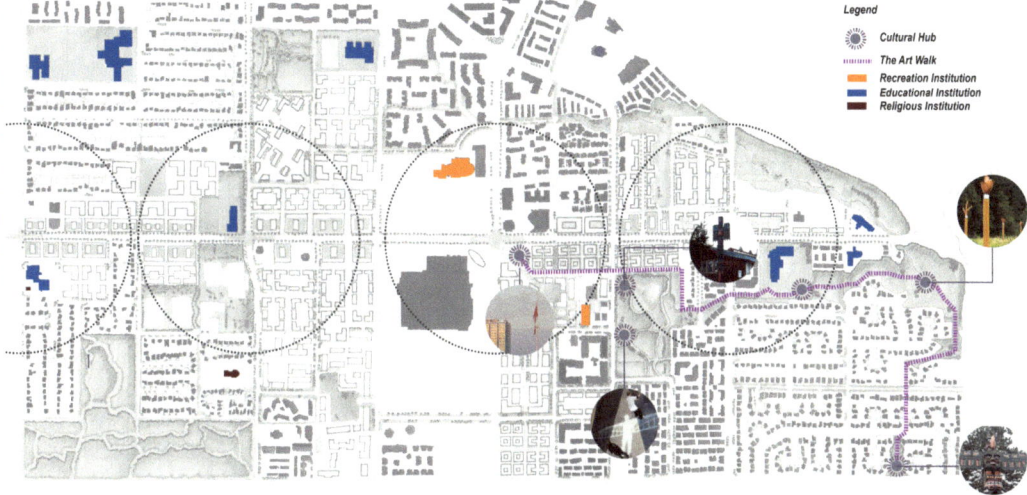

3. First Nation Cultural Route

Legend

- Cultural Hub
- The Art Walk
- Recreation Institution
- Educational Institution
- Religious Institution

4. Core Cultural Area: The string of "Cultural Pearls" denotes that the bridge could become a signature architectural piece, while also providing a weather protection shelter to transit patrons and pedestrians.

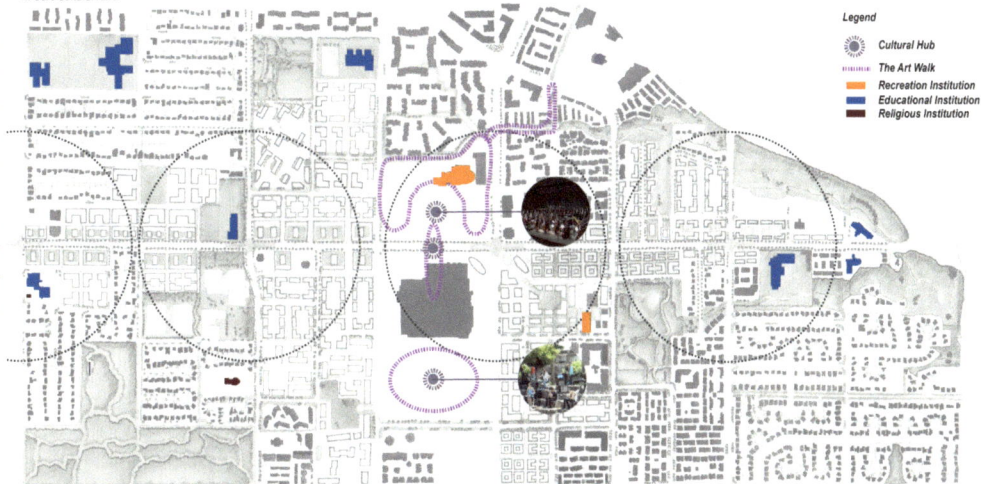

4. Core Cultural Area

Legend

- Cultural Hub
- The Art Walk
- Recreation Institution
- Educational Institution
- Religious Institution

Road Network

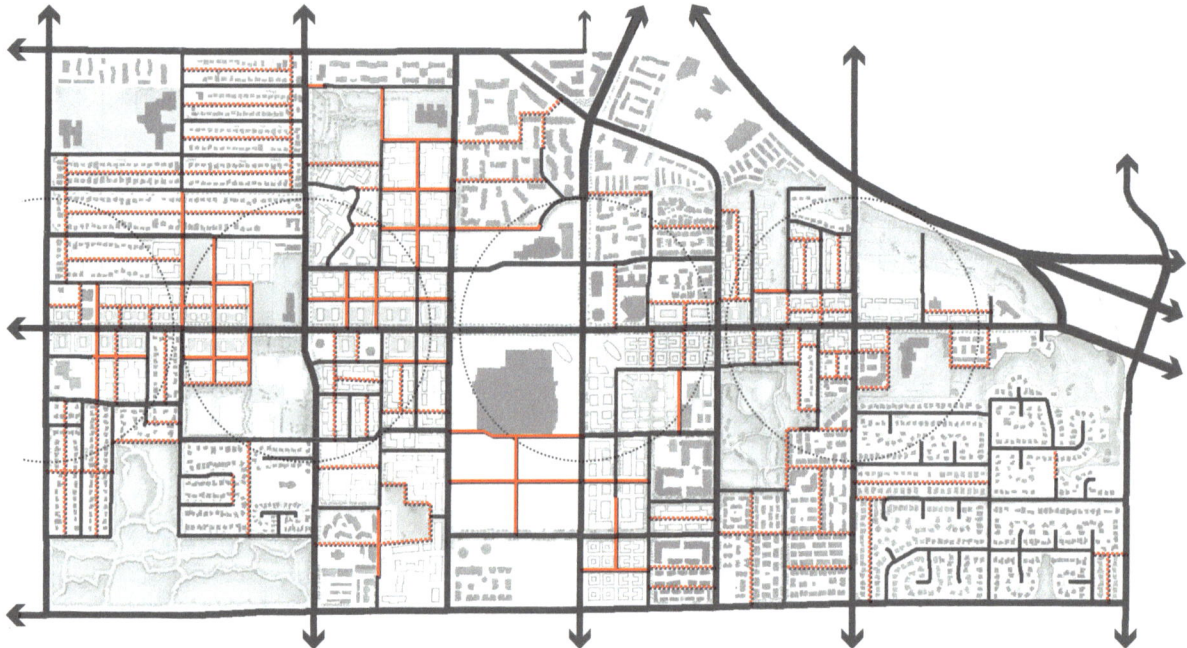

LEGEND

━━━ Existing Streets

━━━ Proposed Streets

〰〰〰 Proposed Access Paths

To transform the existing car-oriented pattern of the site for greater connectivity, the road network, as the essential element of the movement system, is carefully examined and improved to become more interconnected in the scheme. New roads and paths (including back lanes) are proposed to increase the local permeability and walkability (as highlighted in red in the diagram above). The reduced site frontage lengths will respect neighbourhood scale. Moreover, the new proposed streets and paths will directly connect to 104 Ave. and will enhance economic value through increased developable frontage in the transit-oriented, intensive development corridor.

Lilian Zhang

Comprehensive Movement System

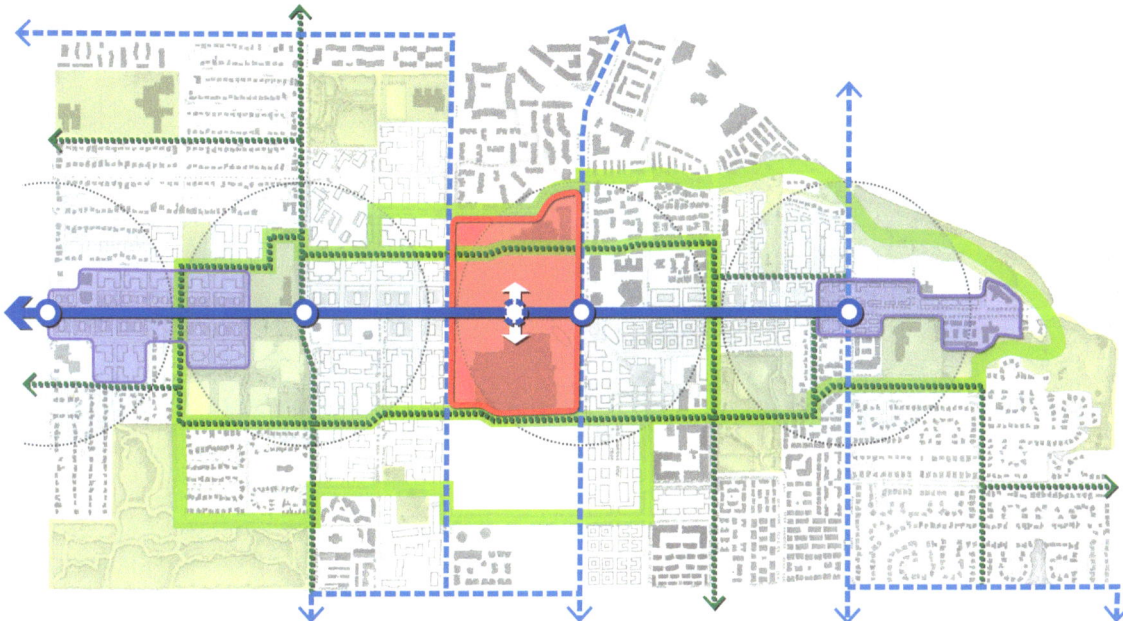

LEGEND

- ⊙—◯ LRT Line & Stops
- ←--→ Bus Routes
- ◀••••▶ Greenways & Bikeways
- ▱ Green Loop
- ▢ Parks & Sports Fields
- ▢ The Heart
- ▢ Welcome Mats
- ⬡ Alternatively LRT Stop

The new LRT line running along 104 Ave. will be the spine of the comprehensive movement system leading the site towards a more low-carbon, transit-oriented developing mode. The rapid transit line, supplemented by the local bus routes proposed in the scheme, will cover most of the neighbourhoods in this area within 5 minute walking distance from transit stops.

There are three development nodes around LRT stops proposed in the scheme: two mixed-use welcome mats and a "Heart"—the New Guildford Center regenerated from the existing mall. An alternative LRT stop is suggested to be located under the existing skywalk of the mall so as to take full advantage of the bridge. It will also provide an extra pedestrian connection between the south and north sides of the Heart.

Parks, sports fields and urban forests in the site will be linked together with a "Green Loop" in the scheme. New local greenways and bikeways will be integrated into the Green Loop to encourage healthy travel modes.

94

Guildford's Identity--Scale

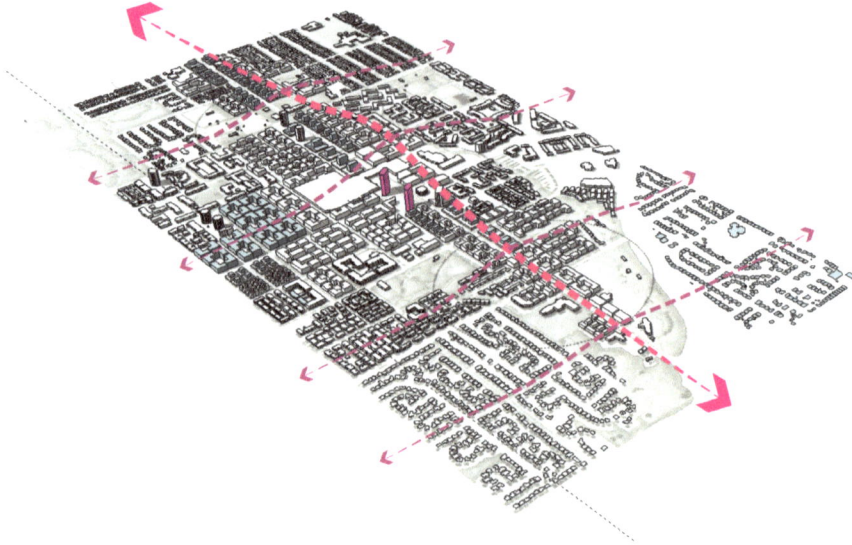

Figure 1: Ascending trend from the Welcome Mats to the Heart (showed by 3D model).

From the model, we can see the ascending mid-rise buildings reach up towards the twin office buildings. There are respective scales between the Welcome Mats (smaller) and the Heart (larger). This trend not only exists along the 104 Ave. transit corridor, but also exists along the north-south orientation streets.

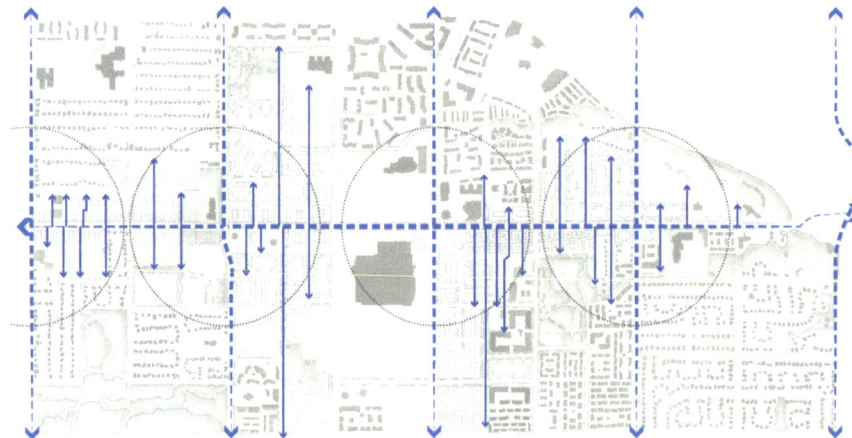

Figure 2: Respective frontage scale along 104 transit corridor.

Through the master plan, we tried to break down the existing long frontages and improve walkability by introducing more north-south local streets. This will achieve more permeable and visually engaging frontages. Besides, each frontage will need to work hard to further break down frontage length through various architectural expressions. Furthermore, we also propose more north-south orientated, mid-rise structures which will achieve a more open corridor experience while maximizing natural light and good views.

Saki Xueqi Wu

Guildford's Identity--Image

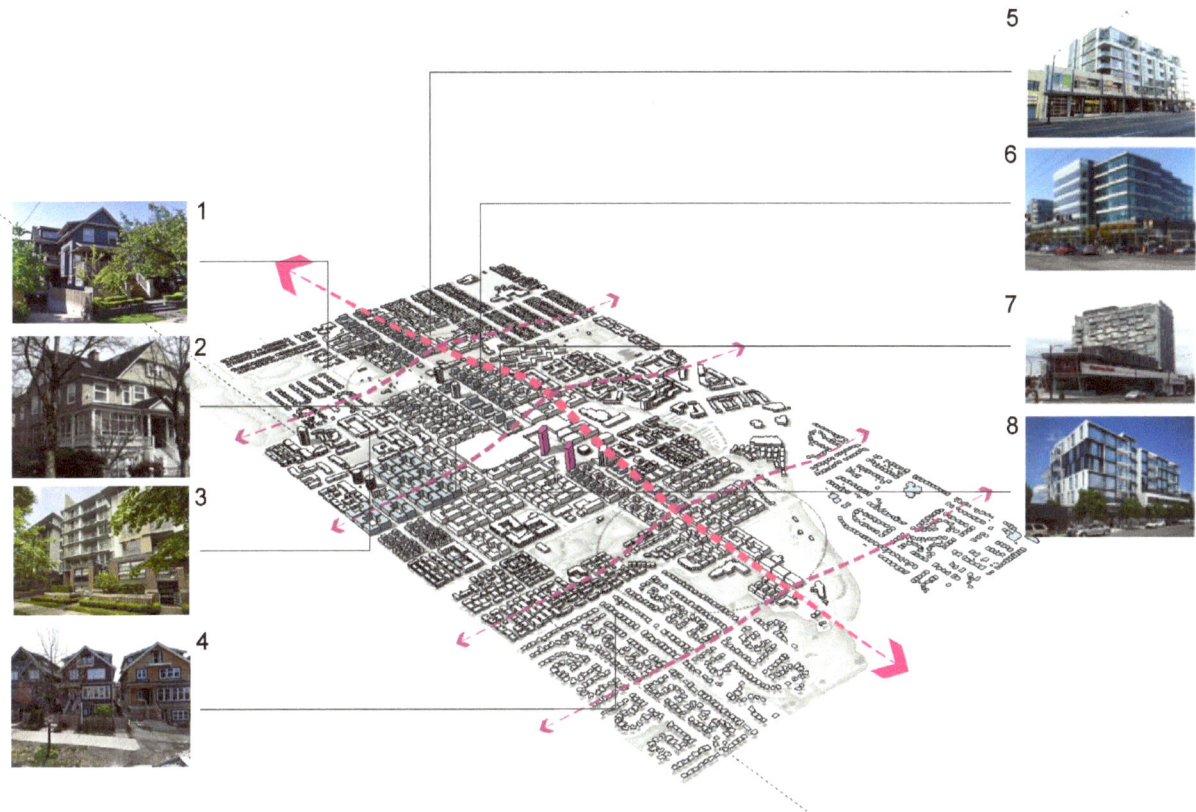

Figure 3: Images of selected housing typologies chosen from the case study catlog for the residential patches of the contextual quilt.

Furthermore, we also selected residential typologies to emphasize on relative scale for the location and housing capacity for each patches of the "quilt":

1. Case6: 1868 W 7th Ave
2. Case28: 2588 Alder St
3. Case48: 2137 W 10th Ave
4. Case2: 1634 Grant St

5. Case60: 2528 Maple St
6. Case65: 507W Broadway
7. Case62: 368 Kingsway
8. Case63: 125 E 8th Ave

Reference: Vancouver Building Typologies List

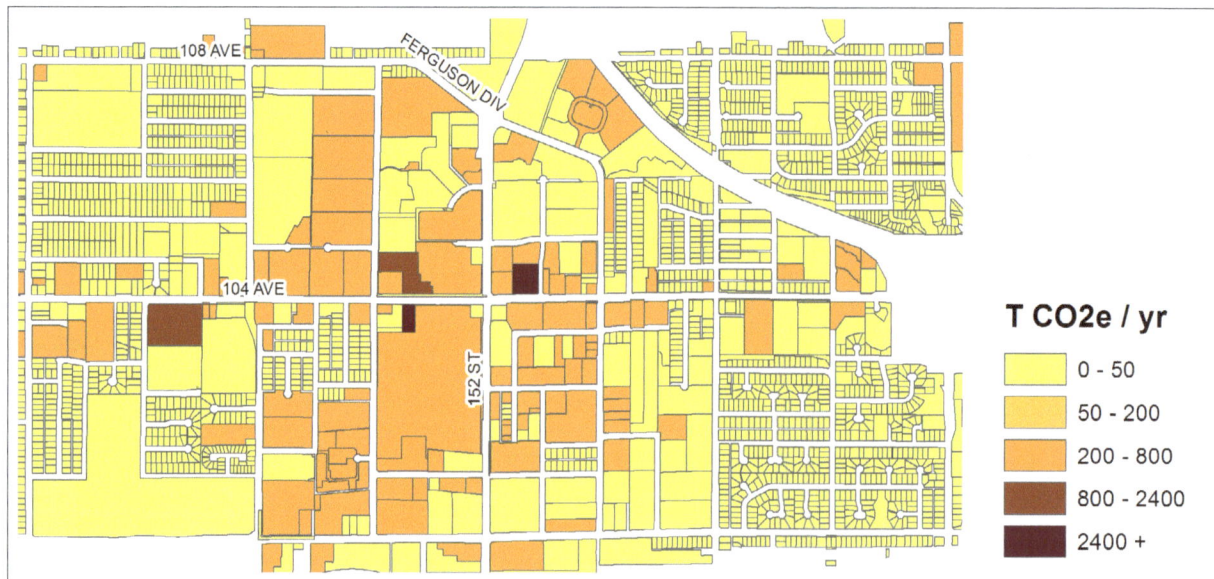

Reference : "Rory Tooke, Community Energy Planner, City of Surrey"

The guidance of Rory Tooke, the community energy planner for the City of Surrey, was instrumental in understanding the energy systems of Guildford. The above diagram illustrates the building emissions for the area of interest: Guildford. In conversation with Rory Tooke, he explained the above schematic. According to the summary made from that discussion, here are the following key factors of the building emmissions analysis.The Energy use intensity (EUI) of buildings is taken from Canadian residential and ICI surveys conducted by NRCAN.The Appropriate EUIs are multiplied by floor area of buildings on each lot. Most of the City of Surrey's building types are aligned with closest match from surveys. The Split in energy sources assessed using same fraction as Community Energy and Emissions Inventory (CEEI) data.The Natural Gas fraction of energy load multiplied by Carbon-equivalent content.The Aggregated results for the entire City of Surrey are within 3% of CEEI measurements. The T CO2 e / yr represents the Tonnes of Carbon Dioxide equivalent per year and the value for Guildford is 40,000 T CO2e / year.

Yashas Hegde

Legend

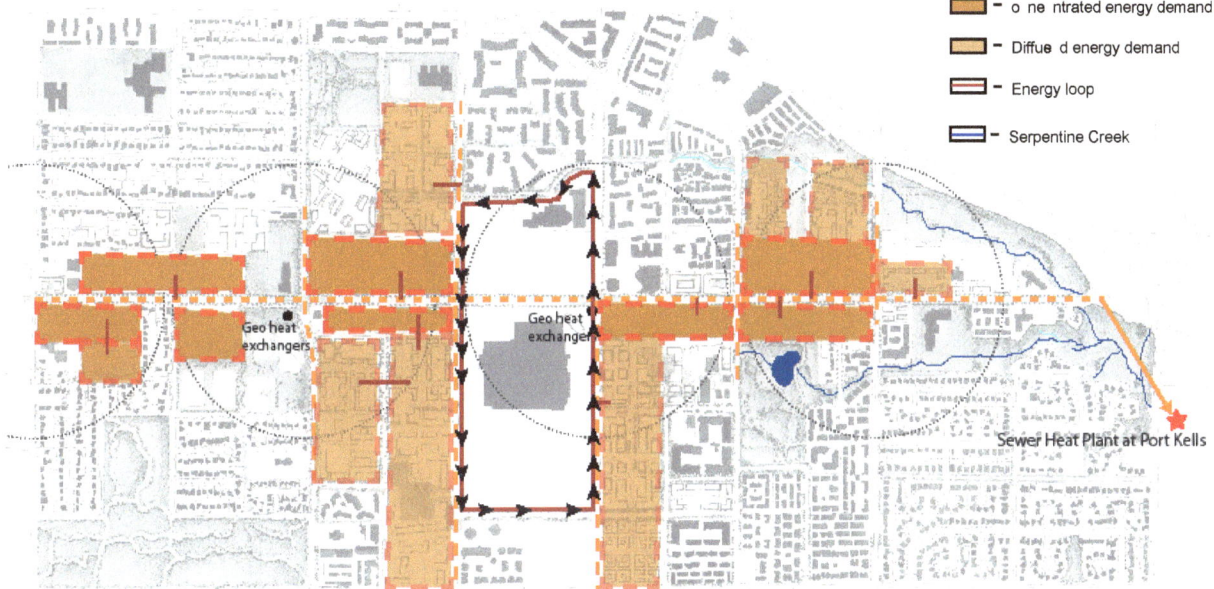

- ☐ = o ne ntrated energy demand
- ☐ = Diffue d energy demand
- ☐ = Energy loop
- ☐ = Serpentine Creek

The above diagram illustrates the conceptual idea for water, energy and movement. The light rail transit investment on 104 Ave. has significant impact on the use of land parcels along the 104 Ave. The basic need to shift from the current typologies to compact urban typologies is an inevitable urban design strategy. Consecutively, this strategy ensures a more energy efficient use of the district energy line along the 104 Ave. It is economically not viable to connect district energy to buildings that are old, hence the district energy lines do not extend to zones where development of land is minimal. Geo-heat exchangers ensure a positive energy dissipation which results in a smarter energy grid. This strategy contributes to an energy loop around the Guildford Mall which shall be dissipated to the surrounding parcels of land which use significantly less energy. The district energy system shall get connected to the Sewer Heat Recovery Plant at Port Kells.

These six diagrams below incrementally layer to combine into one inclusive master plan layout.

Master Plan

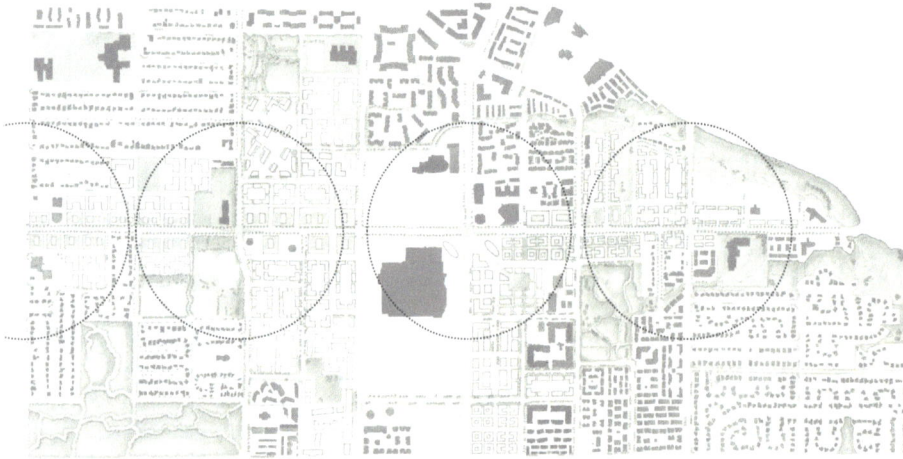

Master Plan + Contextual Q

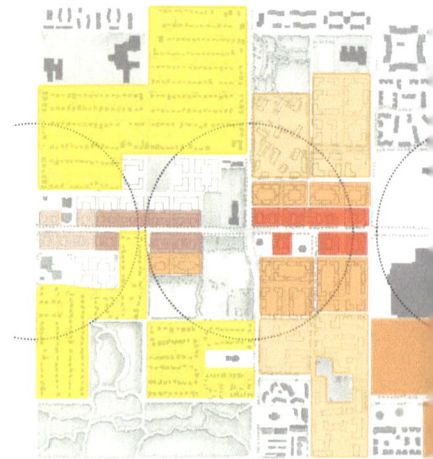

Master Plan + Contextual Quilt + Green Loop + New Streets

Master Plan + Contextual Q + New Streets + Green Way

Weicen Kate Wang

Master Plan + Contextual Quilt + Green Loop

Green Loop

Master Plan + Contextual Quilt + Green Loop + New Streets + Green Ways + Welcome Mats

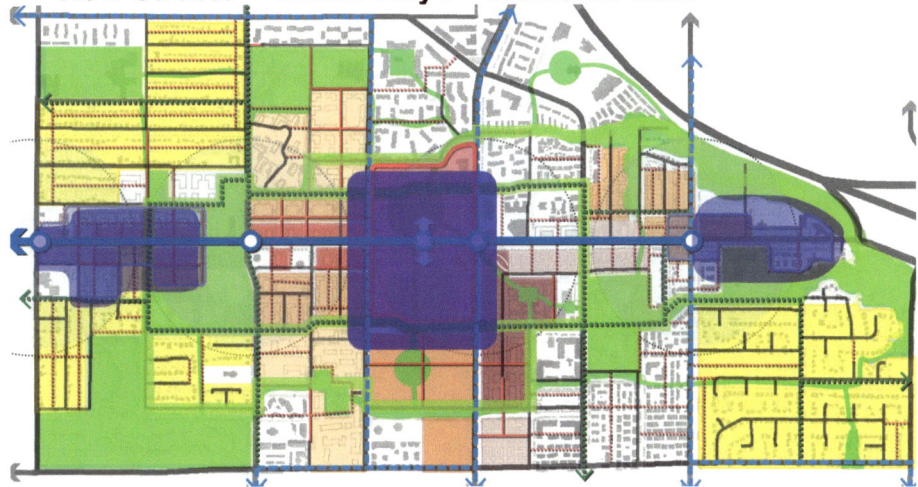

Master Plan + Contextual Quilt + Green Loop + New Streets + Green Ways + Welcome Mats

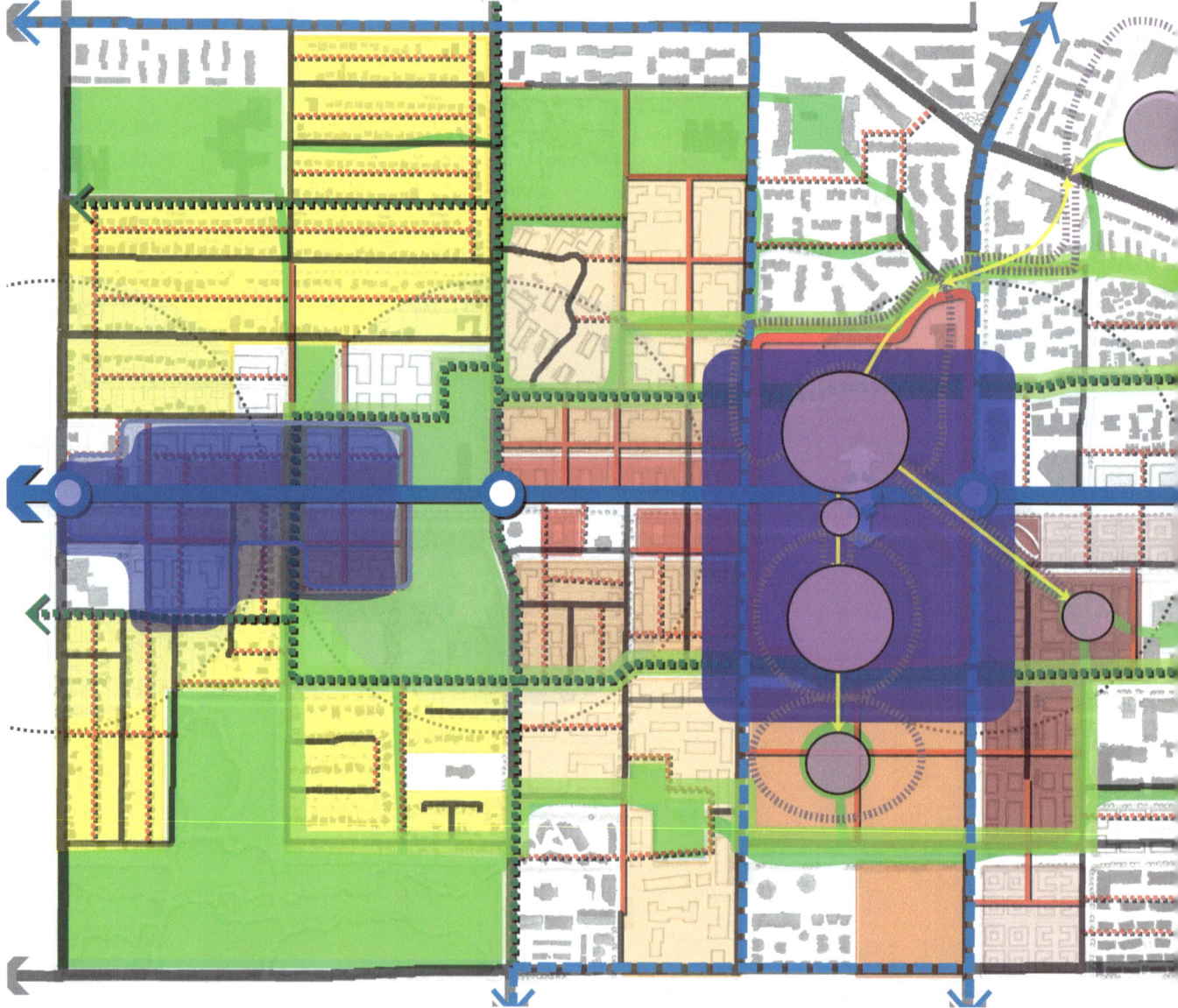

red Pearls

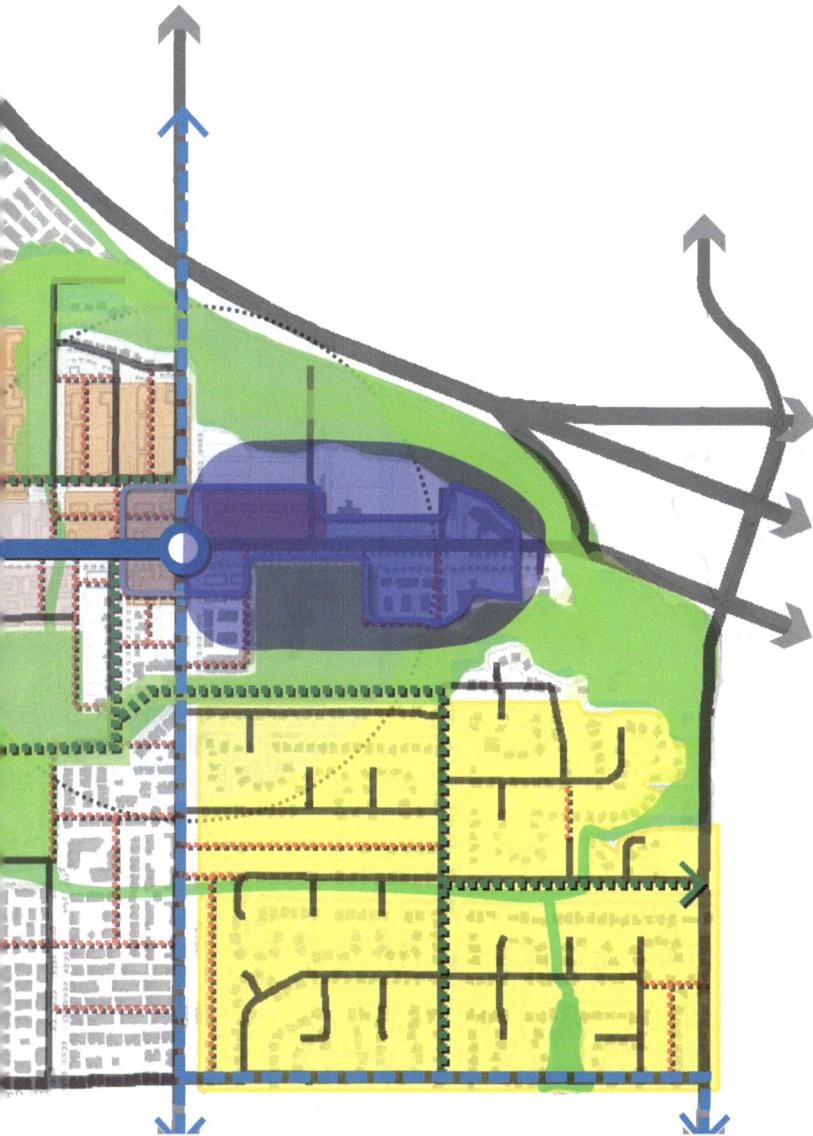

The master plan design includes the Contextual Quilt, the Green Loop, the Movement Systems with New Streets and Greenways, the Humanized Transit Corridors with Welcome Mats and the Heart, and the Cultured Pearls. In conclusion, this design is the result of an integrated urban design framework. At the same time, it is based on the whole systems thinking.

Chapter 6: Framework Investigations

To reveal, and emphasize, the deeper potential embedded in the proposed urban framework for Guildford, each designer produced a concluding vignette. These vignettes investigated specific proposals for urban systems, strategic sites and innovative commercial/residential typologies with each demonstrating the potential for interpretation, and expression, of the unfolding Guildford Community. The studio's overarching proposed urban framework demonstrates, through the vignette investigations, the potential for many and varied "design outcomes" that may, or may not be predictable at this time. This important idea remains central to effective urban design strategies that are always implemented over longer periods of time. The studio's work in Guildford is transferable to any context, culture and scale.

6 FRAMEWORK INVESTIGATIONS

Legend
— Water System
∷∷∷∷∷ Green way
■ Green space
■ Recreation Institution
■ Educational Institution
■ Religious Institution

Fig 1: The green system

Green System: Green Space+ Green way

The Guildford green system serves as a habitat and a recreation corridor by linking the site to the parks and natural forests. The use of neighbourhood grid is to design the green ways that connect the neighbourhood, green space and creek as a whole green system. The green way ensures that cars, bicycles, and pedestrians can always travel through the most direct route. The green system includes several types of green space, parks, street parks, natural forest and community gardens. These different types of green spaces contribute to the neighbourhood's ecological and sustainable development.

The Flood Plan

The flood plan has six different strategies to deal with the rainwater collection: park, urban plaza, street, green street, natural creek, urban creek. The rain water flowing along the neighbourhood grid is necessary to be allowed to flow with a natural and gradual path along the sides of the streets. All neighbourhood stormwater is brought to the rain water collection system, where it is collected in channels and artificial marshes. As more water gets collected, the rain water collection widens the green street decreasing the speed of the flowing rain water. The natural creek also becomes a main recreational amenity and visual landmark for the district.

Xinyun Li

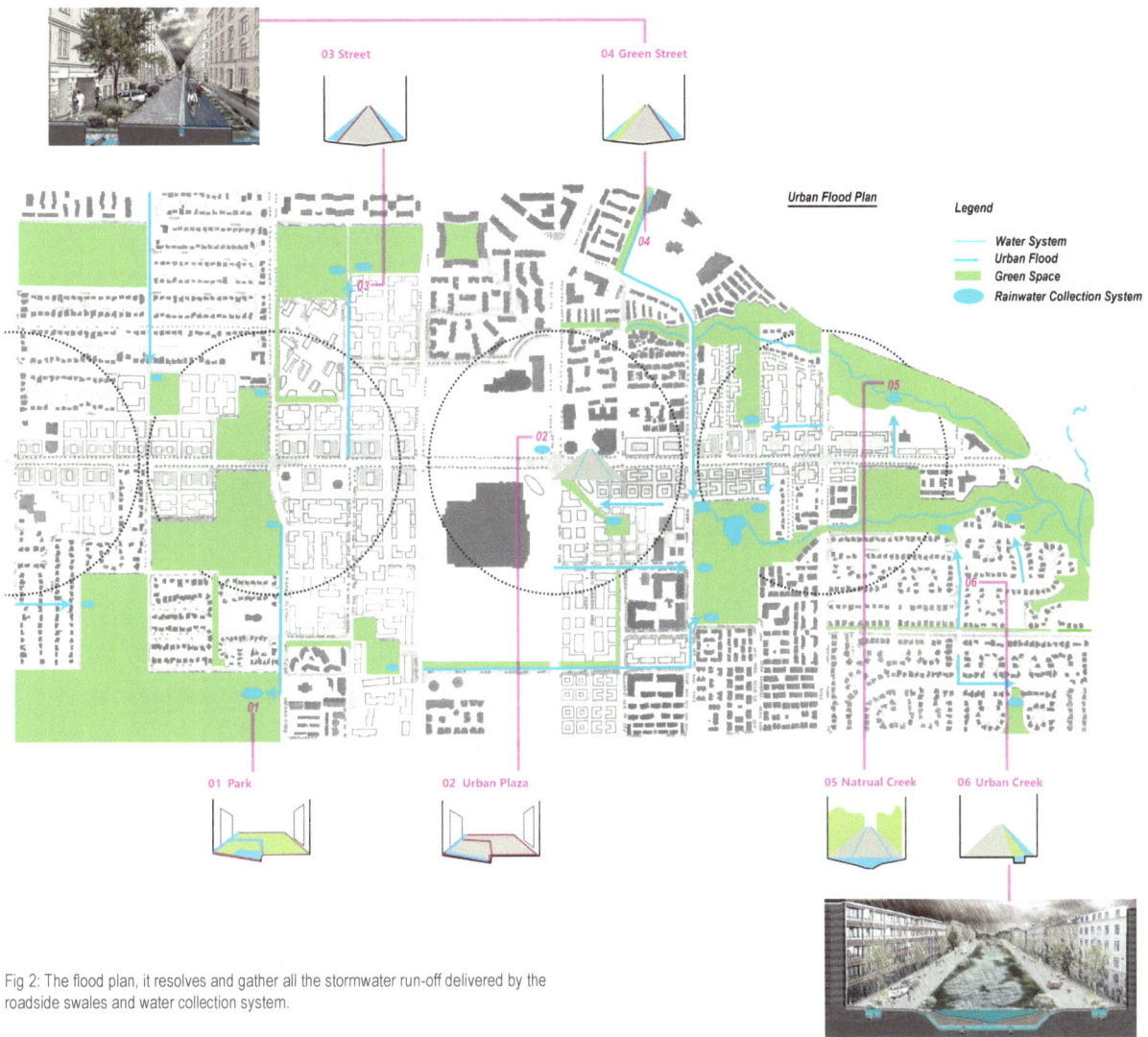

03 Street

04 Green Street

Urban Flood Plan

Legend
— Water System
— Urban Flood
▮ Green Space
⬭ Rainwater Collection System

01 Park

02 Urban Plaza

05 Natrual Creek

06 Urban Creek

Fig 2: The flood plan, it resolves and gather all the stormwater run-off delivered by the roadside swales and water collection system.

108

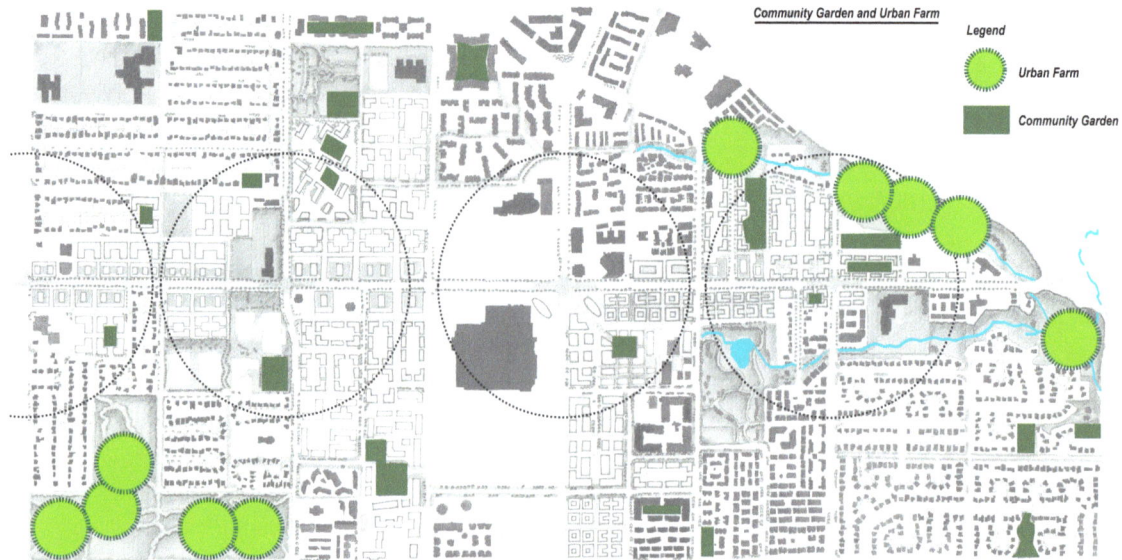

Fig 3: The accessible community garden and urban farm

Community Garden and Urban Farm

Approximately 12% of people in Canada are disabled, so it is significant to make provision for accessing the community garden plots. These lowest incomes and those with extremely low home ownership possibilities are unlikely to have their own gardening space. The community garden can plant local foods to create urban agriculture, however, it is not just a place to grow vegetables. The place has multiple uses, such as a farm, a playground, a sanctuary and a habitat for wildlife. The community garden promotes communication among people and grow stronger engagements when they work together. Community gardens teach and celebrate values to cherish, including cooperation, volunteering, respect for bio-diversity and ecological awareness.The urban farm is a kind of a urban agriculture that includes the process of cultivation, processing and distributing local food in Guildford. The urban farm plays a significant role for growing sustainable communities. The urban farm can be used to grow more organic local food, provide easy access to fresh fruits and fresh vegetables. It can also improve the food security and food safety for the region.

Xinyun Li

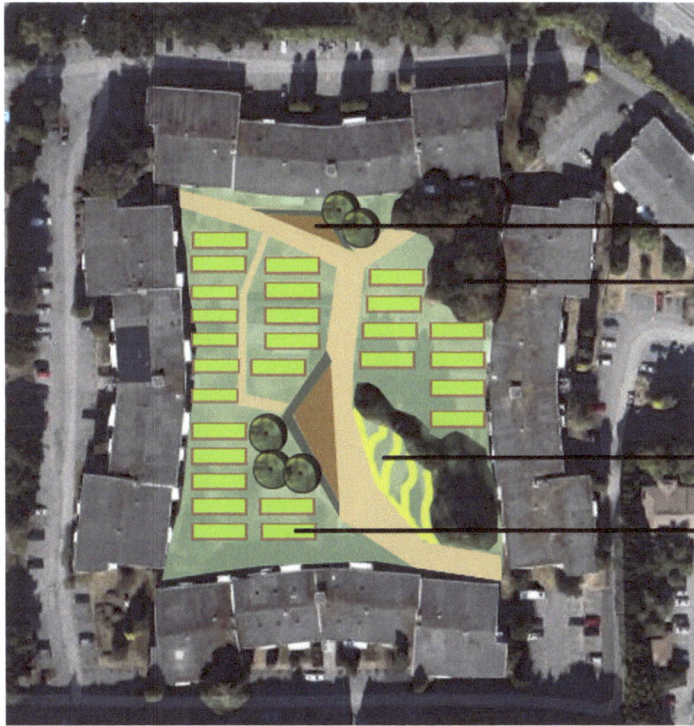

Rest Area

Existing Planting

Decorative Fence

Garden Panel

Fig 4: For instance, the community garden design is not just a garden, but can also be a playground, gym for people to interact and do perform various agriculture-related activities.

110

Green Spaces
Urban Agriculture
Markets
Schools
LRT Stop

104 AVE

5.8 km south
to the ALR

The Food Loop looks at Guildford's proposed green loop through an agricultural lens. New community gardens and urban farms are highlighted among the existing green spaces. Local food markets are also proposed within walking distance of residential areas and in relation to the final stop of the LRT line along 104 Ave. A larger farmers market in the centre of the site can work with neighbouring commercial, cultural and community amenities.

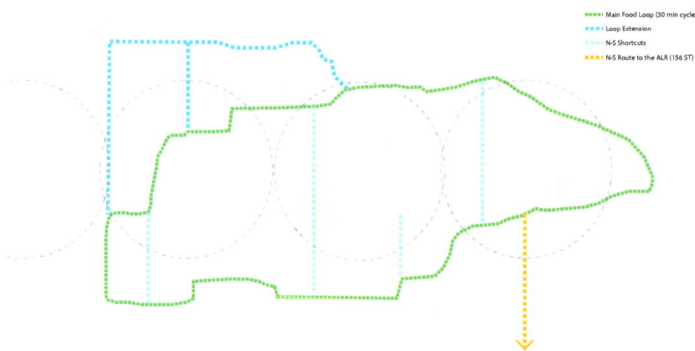

Main Food Loop (30 min cycle)
Loop Extension
N-S Shortcuts
N-S Route to the ALR (156 ST)

20 min cycle
30 min drive
60 min walk

ALR   Guildford Study Area

This diagram shows the proximity of the Agricultural Land Reserve (ALR) to Guildford. 156 St. represents a direct path of travel for produce from local, larger scale farms to the study area. This distance can also be traveled comfortably by bicycle as an extension to the Food Loop bike route.

The Food Loop is envisioned as a bike loop. The main recreational path takes 30 minutes for an experienced rider, but can also provide a day-long activity for families. There are also possible extensions and shortcuts within the network.

Simone Levy

The conceptual site plan for the area northwest of the final LRT stop shows examples for courtyard and podium building typologies that can accommodate urban agriculture. It also features an urban farm neighbouring the riparian green spaces surrounding Serpentine Creek. This district can become an area showcasing local, sustainable farming strategies at varying scales, and can link the final stop of the LRT line to Surrey's rich agricultural identity. It can also become a place of local produce distribution that is grown on site or in the more robust, nearby ALR.

Urban Farm

Typical Courtyard

Podium

Mixed-Use Building With Local Food Market

156 ST

104 AVE

Shared Farming

Private Agriculture

Native Pollinator Buffers

Courtyard Strategy Replicated

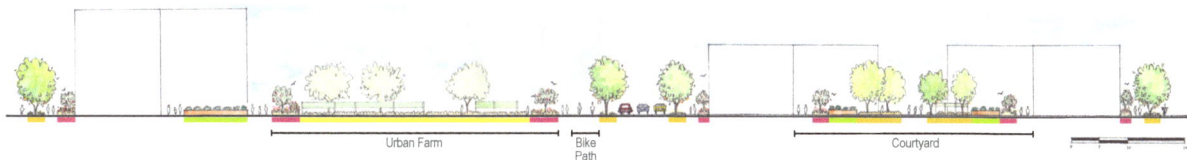

Urban Farm

Bike Path

Courtyard

This section shows the relationship between the new multi-family residential buildings, public urban farm, bike path, and agricultural courtyard. The streetscaping features edible plants, including fruit trees, in order to mark this area as a food-focused district.

Courtyards
Podiums
Single-Family

Garden Plots for Building Residents

Communal Agriculture Space

Native Pollinator Habitat

Three different strategies for small-scale, productive urban agriculture are proposed across the Guildford study area: (1) Courtyards, (2) Mid-rise Podium Rooftops, and (3) Single-Family Shared Yards. The prevalence of these strategies across the site can be seen in the diagram above.

Courtyard buildings are used across Guildford to provide highly livable multi-family housing with access to open space. A strategy for designing these courtyards to allow farming at multiple scales allows for the production of local food and opportunities for community building.

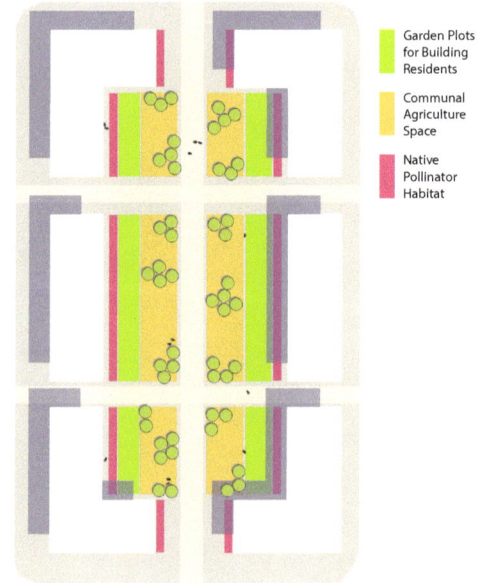

Courtyard Strategy

Communal agriculture areas can be used for farming practices requiring more space, and can be shared among local residents. Possible uses for these spaces include orchards, vineyards, and shared chicken coops.

Private-scale, more intensive garden plots for residents of the buildings surrounding the courtyards. A diversity of produce crops can be grown for each household.

Native pollinator hedgerows. These provide buffers between agriculture and natural green spaces, and provide habitat for pollinators that improve the health of the agriculture.

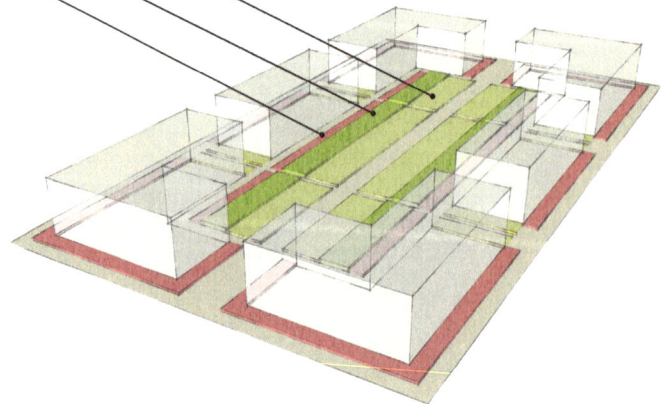

Simone Levy

Higher
Intensity
Farming

Communal
Open
Space

Native
Pollinator
Habitat

Midrise
Structure
Roof Planted
With Natives

104 AVE

Podium Strategy

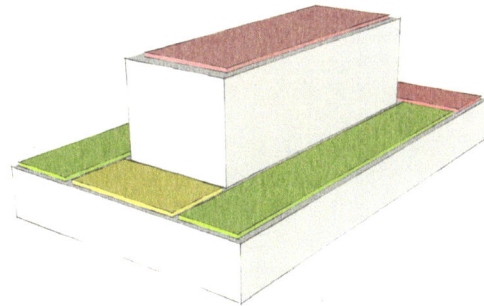

Similar to the proposed courtyard strategy, mid-rise podium structures can include multiple forms of agriculture. Higher intensity farming beds and communal open space are accessible to residents, with drought-tolerant pollinator habitat at the highest rooftops. These habitat patches are not accessible to people and help improve the biodiversity value of the surrounding neighbourhood.

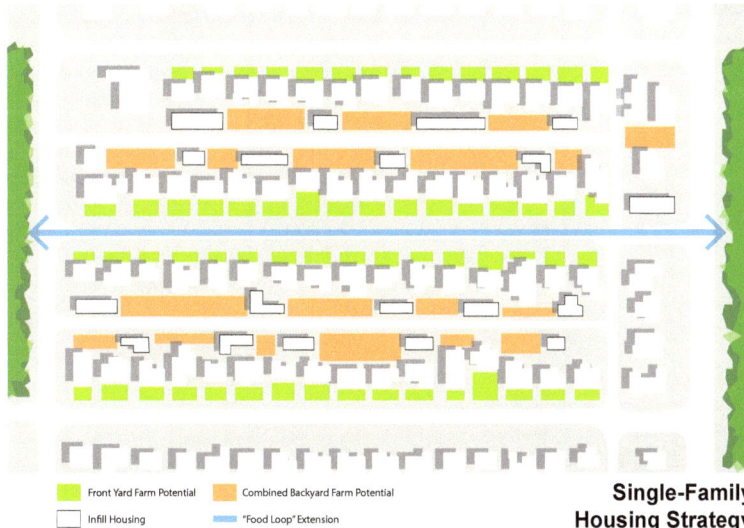

Front Yard Farm Potential   Combined Backyard Farm Potential

Infill Housing   "Food Loop" Extension

**Single-Family
Housing Strategy**

In preserved single-family neighbourhoods, back yards and front yards can be shared between neighbours in order to create larger, and more productive spaces for local farming. This strategy works along with proposed sensitive infill housing to create a more diverse character within these residential areas.

Introduction

Agriculture is one of Surrey's many identities. Surrey's 2013 Official Community Plan (OCP) argues that urban agriculture is able to "resolve rural-urban conflicts and provide healthy and affordable food" (Surrey OCP 2013). The plan also suggests that Surrey should promote a safe and convenient transportation network and new development opportunities that support agriculture, economy, and communities. This project tests methods of how to integrate locally supported agriculture into existing neighbourhoods and potential new urban developments. The project site is bounded by the 154 St., 156 St., 100 Ave., and 104 Ave. (Figure 1). It is located on an edge that transitions from an urban context to rural condition. The key strategy of this project is to sustain both the urban characteristics of varied economic activities and develop a sense of suburban village, nature, and visual aesthetics.

Figure 1: Site location. Source: Google Map

Existing Conditions

The site is comprised of several pieces of residential areas, a natural landscape with urban forest and a creek, two car dealers along the 104 Ave., and a large unutilized green space (Figures 2 and 3). There are some constraints in terms of applying the idea of agriculture. First, the existing building density is so high that there is not much vacant space left to locate agriculture (Figures 4 and 5). Second, the existing street network is automobile oriented, with poor workability, connectivity, and safety for pedestrians (Figure 6). Thirdly, the edge between the site and the 104 Ave. is not active. However, there are a few economic activities and interactions on the street (Figure 1, 3 and 4).

Figure 2: Site's aerial map

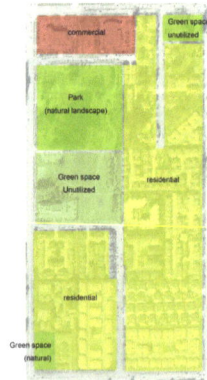

Figure 3: Existing land uses

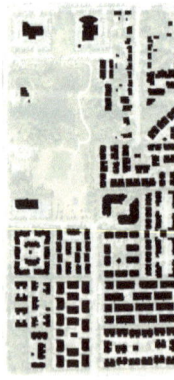

Figure 4: Existing building footprint

Figure 5: Existing vacant space

Figure 6: Existing street network

115

Chen Fan

Design Concept

According to the analysis, the following design concepts were considered:

1. Developing agricultural opportunities on the site and thus creating a unique and memorable sense of place for this community;

2. Promoting a pedestrain-oriented street network that makes this site more walkable and accessible for pedestrians (Figure 7);

3. Encouraging mixed use development, such as grocery stores and seasonal farmers markets on the edge of the 104 Ave. (Figures 8 and 9); and

4. Adding public realms that could catalyze economic activities and social interactions. In particular, they can be used as celebration spaces whenever there are events such as harvesting, fruit selling and outdoor cooking (Figure 8).

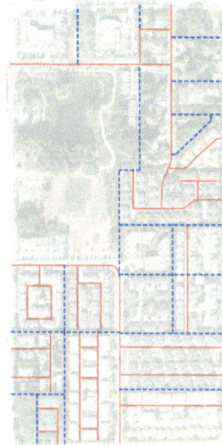

—— Existing street
---- New pedestrian path

Figure 7: Proposed pedestrian network

■ Mixed-use
■ Public realm

Figure 8: Proposed mixed-use development and public realms

■ Existing buildings
■ Proposed buildings

Figure 9: Proposed buildings

Five Strategies of Agriculture

The agriculture opportunity is undertaken based on the inventory of land availability for the given site, including public lands such as parks, public spaces, rights-of-way, and rooftops, as well as private lands such as courtyards of multi-family housing and backyards of single-family housing. These lands could potentially be used for urban agriculture. The proposal develops five specific strategies of urban agriculture and selects optimal locations respectively. The five strategies include: 1. Private homestead gardens in single-family housing; 2. Semi-public community gardens in the courtyard of multi-family housing; 3. Agriculture on the flat rooftop of buildings with concrete structures; 4. Centralised urban farms; and 5. On-street, small scale agriculture.

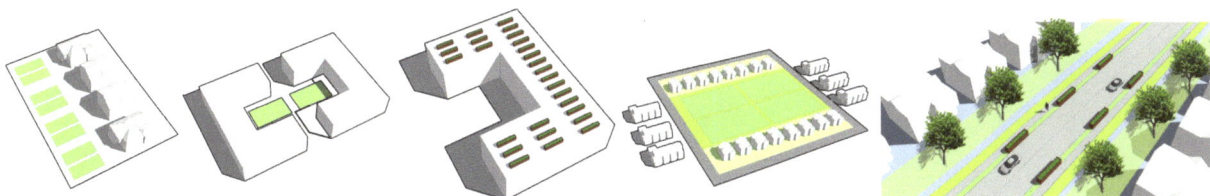

Figure 10: Five Strategies of agriculture: single-family homestead, multi-family court yard, on roof top, centralised urban farm, and on-street small agriculture (from left to right).

Concept Plan

The concept plan is synthesised by the locations of the five agriculture strategies, public realms, natural assets, new development opportunities, and transportation system (Figures 11 and 12). Two sub areas (site A and site B) are designed in detail in the next section.

New buildings

On-street agriculture

Homestead garden

Centralised urban farm

Rooftop garden

Courtyard garden

New street network

Mixed-use program
and public realm

Site A

Site B

homestead   courtyard   on street
rooftop   centralised

Figure 11: The idea of "community agriculture + walkable neighbourhood + vibrant street + public realm"

Figure 12: Locations of the five agriculture strategies.

Chen Fan

Detailed Design: Site A

In Vancouver, most crops require six hours of direct light (Vancouver Community Agriculture Network 2008). This generally means good southern exposure, so agriculture sites must be carefully planned if there are tall trees or large buildings along the south end. A sunlight analysis is done first in order to observe the site in the morning and afternoon and determine which areas could receive adequate sunlight (Figure 13).

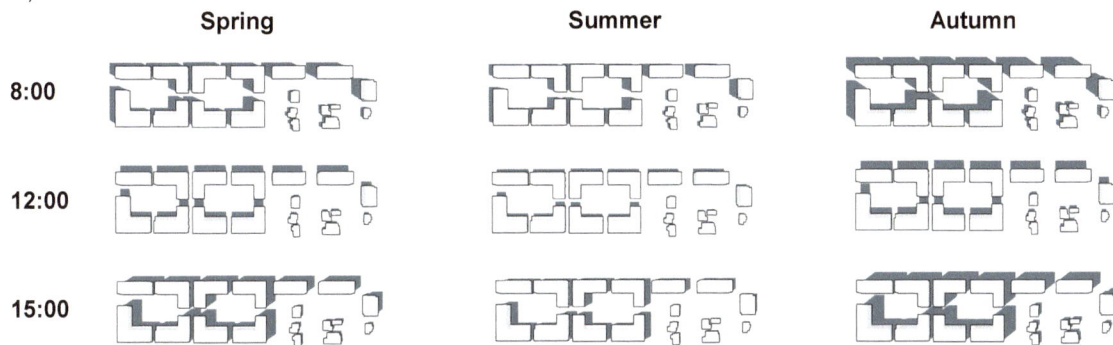

	Spring	Summer	Autumn
8:00			
12:00			
15:00			

Figure 13. Sunlight analysis.

Figure 14 illustrates the final master plan of Site A. All the buildings are five stories high. The first floor is proposed to be commercial while the other four floors on top are residential. The buildings on the bottom of the plan have podiums on the third floor. The containers in the courtyards are located based on the shade analysis above. The spaces between buildings and all the paths add permeability to the site. Taking advantage of the light rail transit (LRT) station, a small urban plaza is proposed to the northeast corner of the site. More design characters are shown in Figure 15.

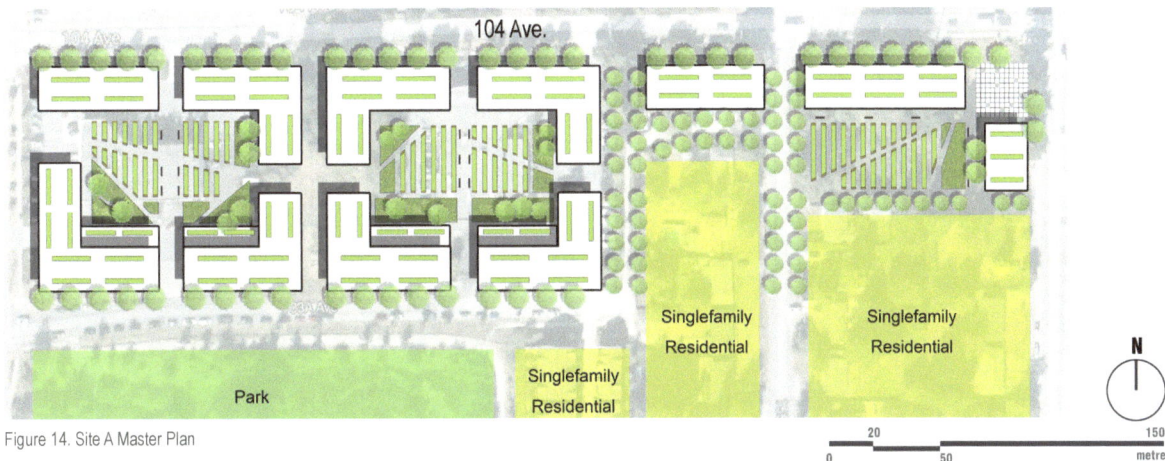

Figure 14. Site A Master Plan

Green spaces inside courtyard
provide opportunities for semi-public activity

Podiums provide space for rooftop garden
and the ground could receive more sunlight

Open corner
adds visibility and permeability
to the court yard

Corner plaza
for relaxing and gathering

Four metres wide continuous setback
provides safe walking & biking opportunities

On-street commercial developments
(retail, service, and fruit/vegetable market)
provide vibrant economic activities and social interactions

Spaces between plant containers
as pedestrian paths

Mid-block paths
provide block permeability and accessibility

Continuous canopies provide
weather protections and pedestrian amenity

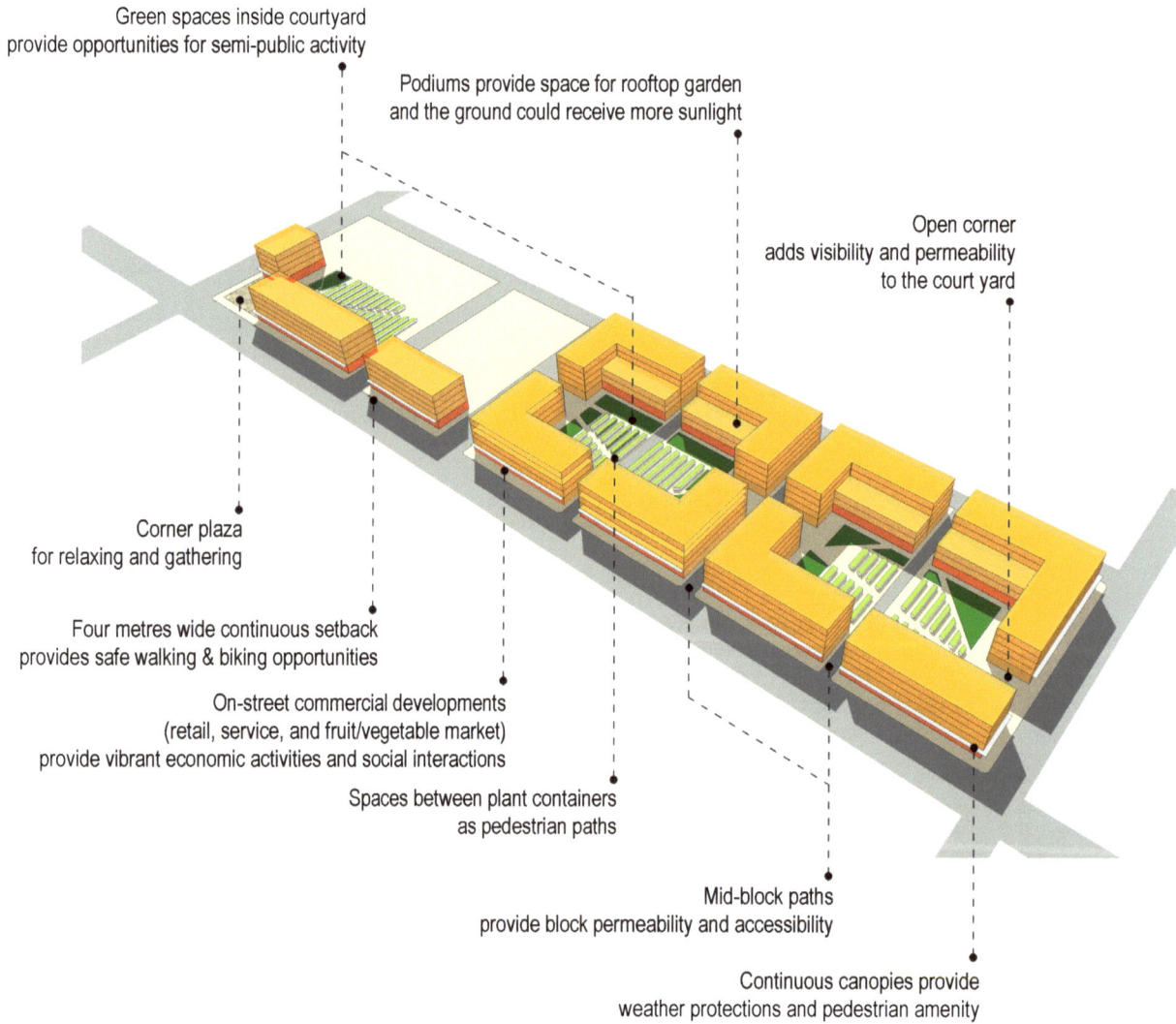

Figure 15: Design characters of Site A

Chen Fan

Detailed Design: Site B

This unutilized space is redesigned to be an agriculture park (Figure 16). The spatial pattern and circulation network are designed based on existing paths. A variety of crops are planted here, which create job opportunities such as maintenance, management and technology. There are some public lands on the south side and the west side of this site. They can be used as parks, ochards and vegetable markets, and celebration space. Moreover, the existing building on the southwest corner of the site is redeveloped as two mixed use buildings. There are some restaurants and retail on the first floor while the other two floors on top are residential. In addition, the site is connected with the existing natural park to the north. Together they generate more possibilities for economic activities, recreational opportunities, and social interactions.

Figure 16: Site B Master Plan

```
      20                            150
0            50                   metre
```

New System: Food+Nature+Community

Figure 17 shows that the two sites, the natural park, the new pedestrian system, and the transit-oriented 104 Ave. corridor integrates into a productive, pedestrian-friendly, green and accessible system. On one hand, agriculture and relavent lifestyles create a sense of village. On the other hand, new commercial opportunities and potential informal food markets sustain certain economic activities. They together build an unique sense of place that would attract more citizens and visitors.

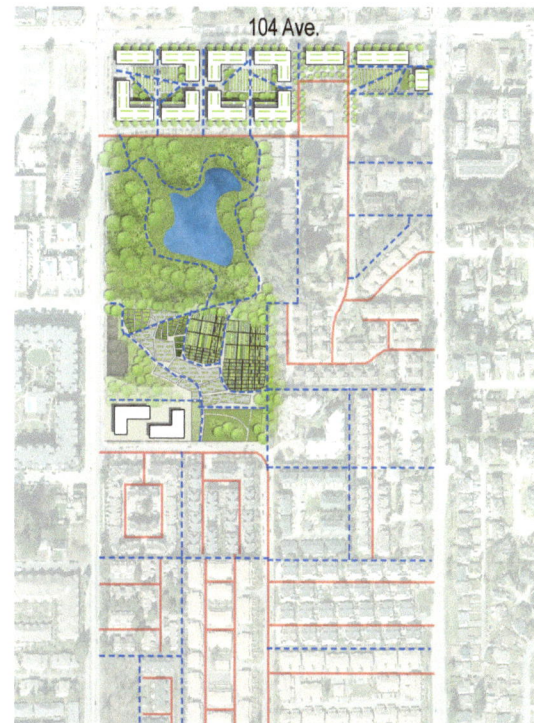

104 Ave.

Figure 17: Final integration illustrates the concept of "food + nature + community

Reference
City of Surrey. "Official Community Plan" Adopted Oct. 20, 2014.
Vancouver Community Agriculture Network. "Growing Community Gardens: a guide to farming food in Vancouver." 2008

The study area is located one block towards the north side of 104 Ave. and the south part of Holy Lane Park on 148 Ave. (Fig1). The site is a gated community segregated from its residential neighbours, local school and park. The dominant structural character of the site is two-storey townhouses (total of 186 dwelling units) with plenty of surface parking, mostly built in 1980 (Fig 2).

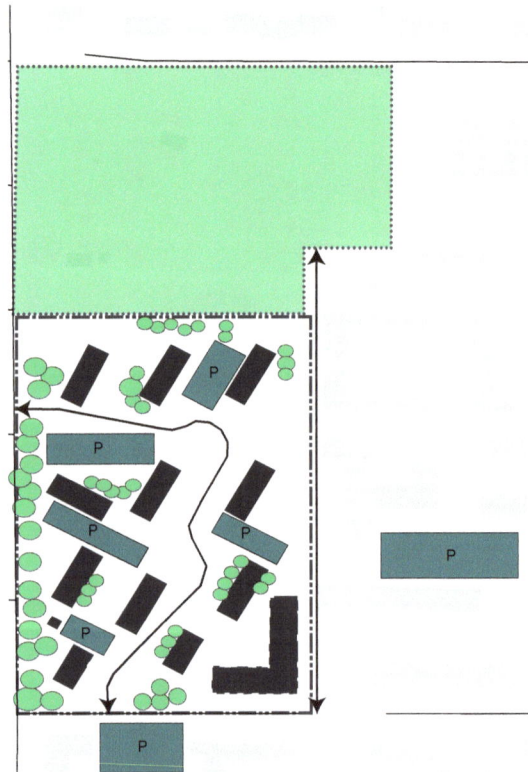

Fig 1: Site Location

Due to the site's closeness to community amenities such as park, school, church, and city centre, the site has a potential to be considered for housing refreshment (Fig 3). In addition, having proposed the LRT stop close to the site (at the intersection of 104 Ave. and 148 Ave.) expands the practicality of the plan in the real world. Transit will boost the market and encourage redevelopment in the area.

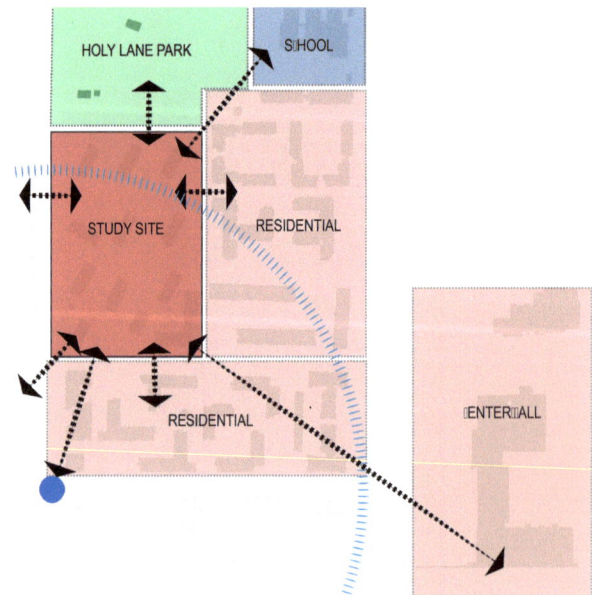

Fig 2: Existing building, tree and parking footprints

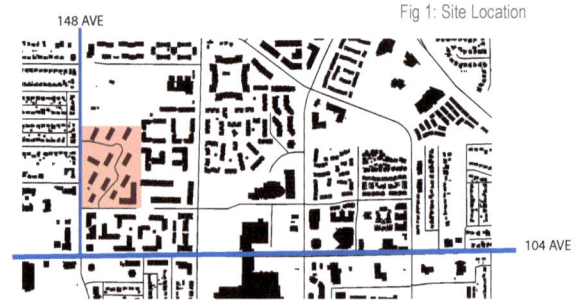

Fig 3: The site potential to be more integrated and connected with the rest of the neighborhood

Hedieh Rashidi

The design proposal aims to create a more integrated community with higher density with respect to the existing conditions.

Four strategies are considered as following:
1. Retrieving the site's connectivity (Fig 4)
2. Keeping the footprint of the building, while adding density
3. Retaining the existing trees
4. Phasing the construction in order to decant the residents on the site and make the development process smoothly

Existing pathways
Proposed pathways
Proposed Public Space

Fig 4: Retrieving the site's connectivity by adding new shared pathways for walking, biking, and local vehicles. In addition, replacing the surface parkings with public spaces with an emphasis on quality of the public realm as shown in the sketches below.

Fig 5: Conceptual Perspective of the site

Proposed Dwelling Unit: 800 to 900 DU
Site Size: 14.8 acre

Building Typology 1:
4 to 6-storey woodframe buildings
Building Typology 2:
2 to 3-storey stacked townhouses

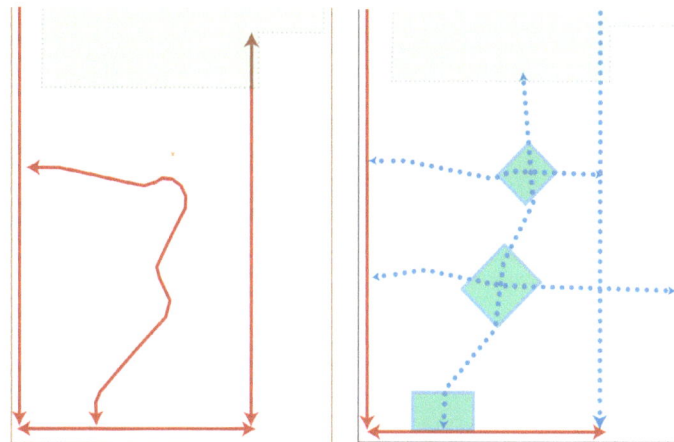

Fig 6 & 7.:Public Spaces

1. Three major public spaces along the south-north pathway, the precedent for these public spaces, shown above.

2. Semi-public open spaces located in the entrance of the buildings with maximum permeability and the visual reference, shown left.
http://imgarcade.com/1/urban-design-concepts-sketches/

Fig 9: First Phase of refreshment

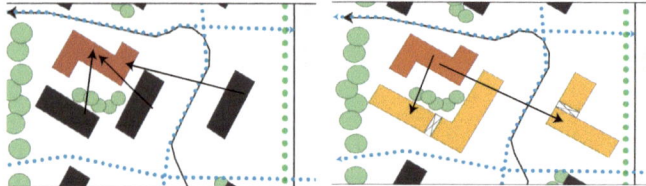

Fig 10: Second Phase of refreshment-Stage one

Fig 11: Second Phase of refreshment-Stage two

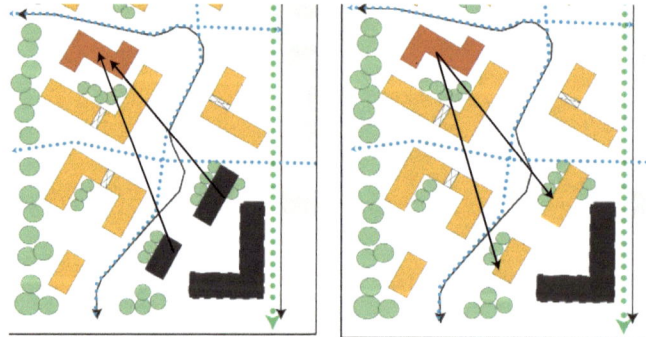

Fig 11: Third Phase of refreshment

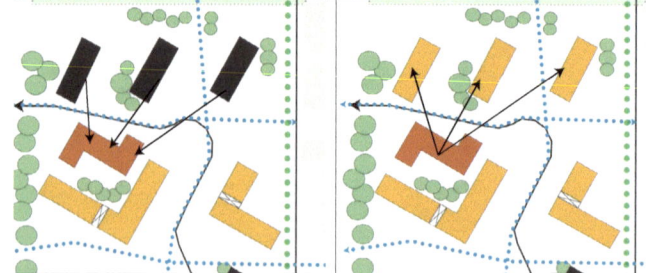

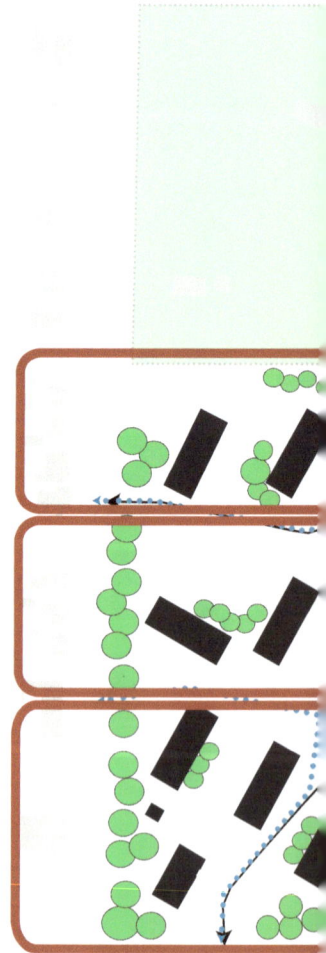

Fig 8: Refreshment of the existing footprints

Hedieh Rashidi

Phase 3

Phase 1

Phase 2

The first phase – two to four year period
The new building (colored in red) is built with maximum density on the former parking lot at the start of site refreshment. The residents are moved to the new building and the other two old buildings are decanted. The newly designed building is located in the same position with an expanded footprint that includes the existing parking lot. The arrows show the residents' movements to the red building and their way back to their newly built apartments.

The second phase – four to six year period
This phase will take place in two separate stages with the same movement as previous construction. The south part of the site will be renovated one at a time. The new buildings in the second stage are primarily a townhouse as it compliments the neighbouring building (colored in black). This will allow the developers to utilize the money generated through community amenity contribution to develop more open spaces on the site (Fig 4).

The third phase – two to four year period
The refereshment project in its final leg will be completed in the third phase. In addition to this redevelopment, special attention has been given to establish a connection with the site and the Holy Lane Park. With this project, the previous gated community is now open to the public. The residents of Guildford and the new immigrant can now enjoy the amenity and live as an integrated community within the larger Guildford region.

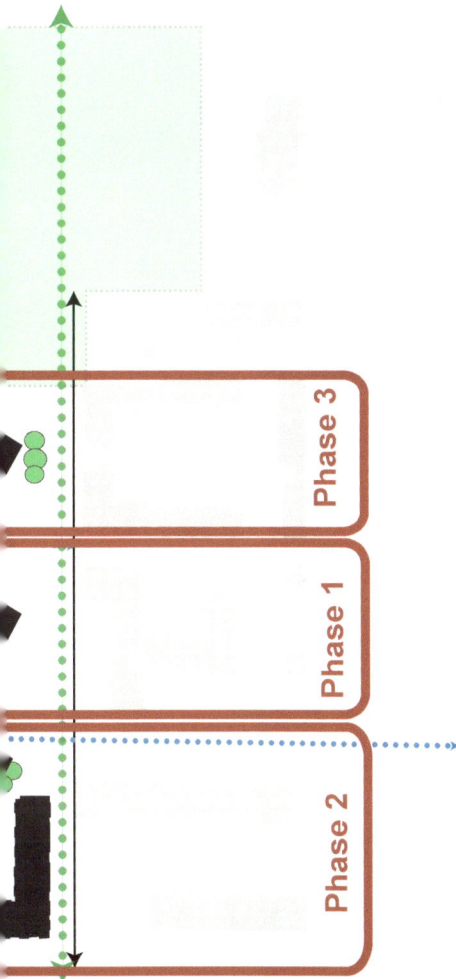

The Urban Fabric of Guildford

The City of Surrey is comprised of six neighbourhoods including Guildford which has an opportunity to become an arrival city due to its strategic location within Metro Vancouver with an additional benefit of Highway 1. A light rail transit is proposed to be developed on 104 Ave. while 144 St. and 160 St. serve as welcome mats to this neighbourhood with Guildford Mall as the heart. The existing fabric allows for additional density which would extrovert the vibrancy of the mall to its surrounding blocks.

Existing streets need to have a refined grid network to enhance the gradual loudness at the heart. The concept of spatial arrangement calls for enlarged spaces, however at the risk losing a sense of cohesion. The entire idea behind Guildford's image is to create a green neighbourhood with optimum utilization of space in relation to human anthropometry. In the end, efficient urban fabric leads to:
(i) Better performance of the city
(ii) A design that saves energy

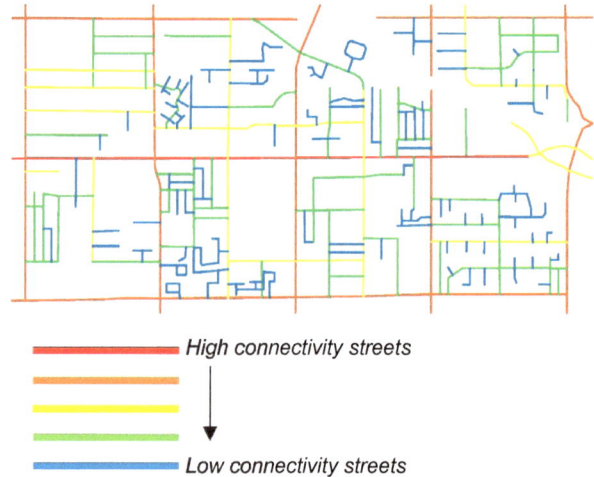

High connectivity streets

Low connectivity streets

Fig. 1: Connectivity of the existing streets which derives the pattern for every block

Fig. 2 shows the proposed structures taking into consideration the volumetric spaces in relation to the height of a building

Prachi Doshi

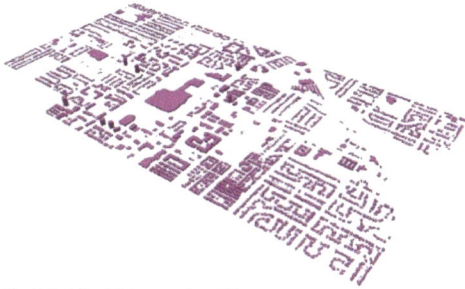
Fig. 3: Building fabric comprises 22%

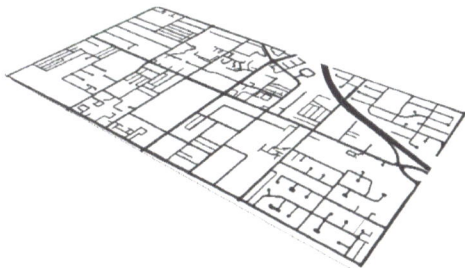
Fig. 4: Paved surface fabric comprises 13%

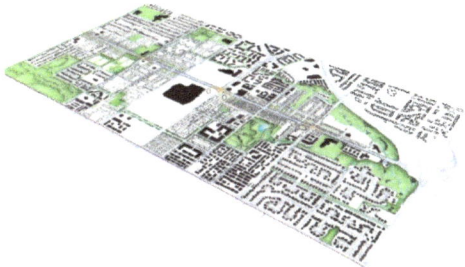
Fig. 5: Green network fabric comprises 25%

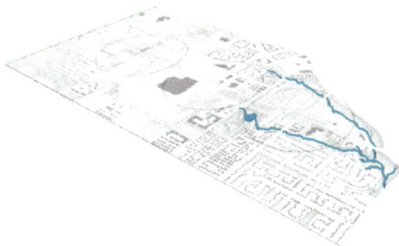
Fig. 6: The stream corridor

Neighbourhood Dynamics

(i)   Systems Thinking – It drives the pattern of these relationships and the way they translate into emergent memberships.
(ii) Complexity – Several possible future scenarios among the dynamics that may exist.
(iii) Self organization – How patterns of relationship internally structured develop over time that needs to be considered.

The interaction of building parcels and parcels of land create a meaningful space for human use. We exhibit a certain threshold limit for proportions and fulfilment of these threshold limits drives our behaviour and leads to a sustainable and resilient community. With the benefit of green areas, Guildford owns land but they do not create a meaningful space. Enlargement loses the sense of enclosure. European cities like Rome are popular for their spatial environment and it is the factor that eventually gives a sense of identity and belonging to the place.

Volumetric Spaces

The stream and the green coverage are vital components contributing to the use of a space. Guildford has the benefit of a green bedspread as a means to connect the varied building typologies. An overall increase in density with a gradual change in building types creates a harmonious diversification.

All the built forms have been extended up to property line to enhance the street life and the podium style structures maintain the streetscape while travelling along 104 Ave. The change in land coverage from the built form is from 22% to 29% with an additional 10,000 units. The built forms have been proposed in accordance to the space created at street level. All the spaces are in proportion between 1:1 and 1:3. Further elongation leads to a sense of street. Space allocation can be a derivative of a combination of conventional and form based codes. The spatial arrangement of these unique elements further allots a meaning to the space. Although volumetric space is the least considered aspect while designing a form, it is the only factor which drives human behaviour and the generated phenomenology as a place.

Gentle Intensification of Single Family Homes

The existing fabric of Guildford is comprised of extensive single family houses. To increase the population, townhouses are proposed on the back lane, which can be constructed by combining more than one property.

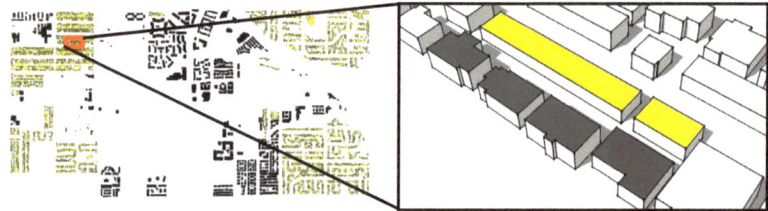

Fig. 7: Gentle intensification of single family homes and a conceptual built form

Multi Family Residential Refresh

Residences have been proposed, which add to the existing low density without disturbing the existing structures. These buildings take into consideration the existing condition and the added structures are in proportion to the space and creates a space between the structures as well.

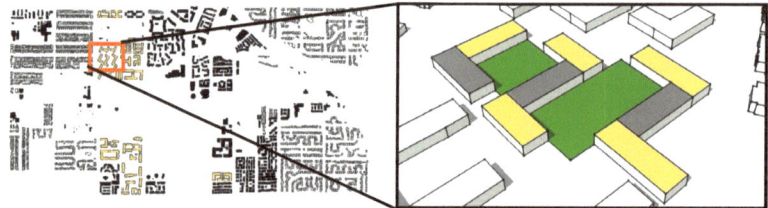

Fig. 8: Addition to the residential areas and a conceptuall built form

New Multi Family Residential 1

New residences have been proposed in the areas away from 104 Avenue and they are four to six-storey structures. Groups of these structures form a space between them, which is used as a courtyard. The south facing buildings have been kept low to cast less shadow in the courtyards.

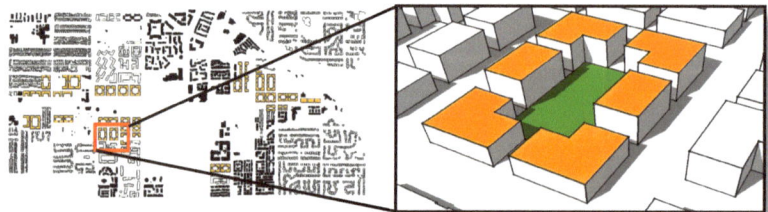

Fig. 9: Medium density residences and a conceptual built form

New Multi Family Residential 2

The south of the mall until 100 Ave. has a potential to be adjacent to the Heart, the Guild and the LRT stop. Hence, six to eight-storey buildings would encourage more people to access the utilities at a walking distance. The building form is a derivative of volumetric spaces and shadows.

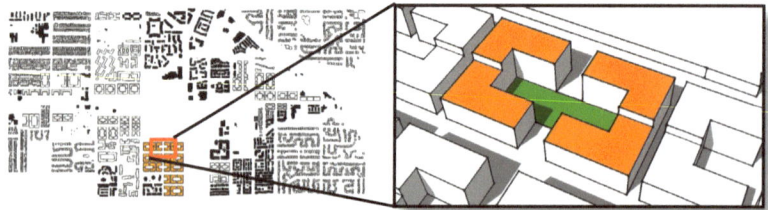

Fig. 10: Higher density residences and a conceptual built form

Prachi Doshi

Mixed Use 1

This category is comprised of the Guild area and at the intersection of 104 Ave. and welcome mats. These buildings are next to the commercial heart of the neighbourhood and provide an opportunity for immigrants. These are four to six-storey structures promoting business corridor.

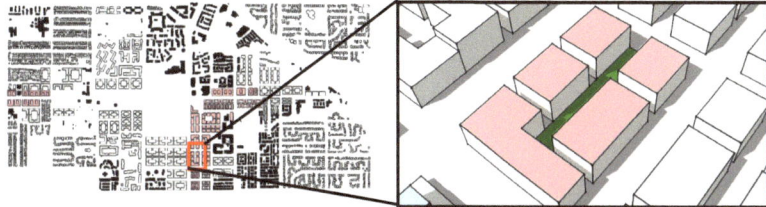

Fig.11: Mixed use 1 buildings and a conceptual built form

Mixed Use 2

These mixed use buildings form a transition from the welcome mats to the core of the neighbourhood. They exhibit podium style typology with seven-storey structures above that and townhouses on the back lane. It provides an optimum use of the space at street level and cast lesser shadow.

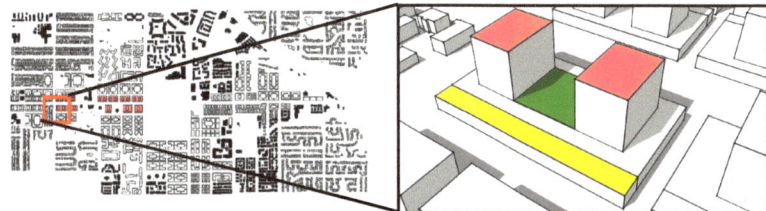

Fig.12: Mixed use 2 buildings and a conceptual built form

Mixed Use 3

Surrounding the Guildford Mall are the mixed use buildings which operate in service to the crescendo at the core. They are two-storey podium with a ten-storey structure above it. They maintain the podium streetscape on 104 Ave. allowing for a higher density set back from the corridor.

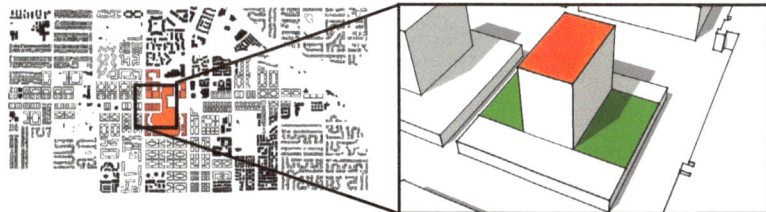

Fig.13: Mixed use 3 buildings and a conceptual built form

Mixed Use 4

Easy access to the LRT stations, the Guild and the Heart leaves room for two towers on 104 Ave. These thirty-storey structures are oriented diagonally so as to cast their thinner shadow on 104 Ave. Such office buildings have the potential to balance the demand for space.

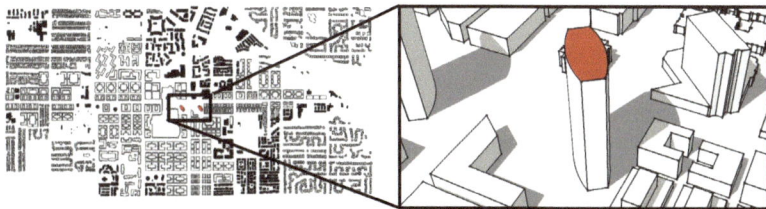

Fig.14: Mixed use 4 buildings and a conceptual built form

Design Goals for the Mall Area:

The design goal is changing the current tendency favoring a large isolated mass over an urban environment towards one generating life and contributing to a vibrant cultural hub. Our main goal is to turn the mall into an urban village with high street experience.

Turning inwards breaking one big mass into smaller mases

Side view of the proposed village showing the podium buildings

A transit-oriented active village is a strategic key component for Guildford Town centre to solve the dispersed suburban sprawl. It would generate a concentration of activities that bonds the area together. It will encourage a walkable heart which brings life to the Guildford centre.

Design Strategies:

-Breaking the one big mass into smaller walkable masses
-Shaping the masses to create hubs and shopping paths
-Opening up the indoor mall to the outdoors
-Replacing the open parking around the mall with underground parking
-Introduce mixed use buildings to intensify land uses

Picture taken of the existing mall from the southeast side

Walkways and Automobiles Paths:
- The Walkway Paths:

Strong access will draw people from the LRT station to a plaza in the middle of the village, directing flow to left and right pathways.

-The Automobile Routes:

The concept limits car movements inside the village to a "U" circle road between the retail and the mixed use buildings, giving car acces to the entire village.

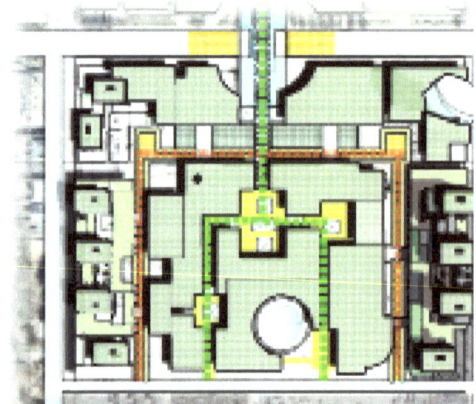

The proposed Walkways (Presented in green color) and Automobiles Paths (Presented in brown color)

Amal Wasfi

In this proposal, the LRT station will be in the middle front of the village. Stairs will take the people from the station to the midpoint of the above bridge which connects this north side to 104 Ave. The bridge have decks enabling people to a wide pleasant view of the arterial supported with cafes.

The bridge configuration is a strong access with glass arches, that orient the people to a plaza in the middle of the village. To allow more than one shopping routes, the plaza opens up to two routes that guide the visitors to the other end of the village.

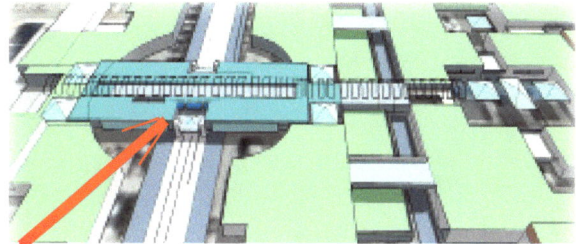

Moving the station to the middle front

The pedestrian paths will come from the middle LRT station

Stairs will take the visitors from the station to the Bridge

130

Green Roofs with flowers and vegetation makes higher level gardens available for all

Land Uses:

Part of the goal is to increase vibrancy and street life. We propose a concentration of commercial and other land uses, where people work, shop, relax, meet friends and also live. These are:

- Mixed use typology with ten residential buildings (800units) above the commercial, adding vibrancy and paying off the cost of the underground parking
- The office tower will provide a concentration of job opportunities in the area and be a landmark for the village.
- Rooftop gardens will control temperature and improve quality of life.

Placemaking:

The aim is to open up the commercial centre to the beautiful natural British Columbia. In the village plazas, we provide advantages of attractive outdoors places, (fresh air, friendly urban environment), flavoured by features of the indoor places, (stairs /escalators, canopies/rain shelters, benches, outside showcases, attractive lighting, artwork, and fountains). Accordingly, spaces are designed to encourage people to move comfortably between the inside and the outside spaces enjoying the coziness. For example, the food courtyard is an open circular plaza connected visually to the below inside and complementing the continuation of the inside.

Stairs and escalators guide people to the plaza effortlessly

The new open food court (at the top of the ground floor)

Amal Wasfi

Creating a vibrant street life:

We are pushing for changes to transform the city from auto zones to pedestrian-friendly zones. This transformation of spaces will welcome people walking and interacting, while making "downtown living possible for a broader range of people."

To increase the integration between the buildings and the different spaces the design works towards:
- Blending the edges between inside/ outside.
- Creating bridges connecting buildings (including residential podiums connected with commercial).

Turning inwards breaking big mass into smaller mases with vibrant street life

- Providing intermittent roof shelters
- Creating defined pathways and plazas
- Providing seating available for anyone's use in an inviting and attractive setting

All of this will contribute to a vibrant urban village to be the heart of Guildford Town Centre.

Outdoor spaces have the feel of the indoor places

Different uses and paths integrated together

Initial Observations:

- Preserve the mall and emphasize its entrances and link them to the external green ways.
- Keep the bridge and remove the Sears building completely so that we bring some light under the picturesque bridge, which already has its unique identity and great potential, but needs some refining.
- Link the existing recreational building to the mall via a piazza that is surrounded by some cultural and mixed use buildings. The residential area will serve the cultural area and vice versa.
- Add two towers that could help the area economically, socially and physically.
- Create volumetric spaces that will lead to the design of the buildings. The design scope will be more likely Phase One of the design. It will more likely be an intervention that is sensitive and selective of what to remove, adjust and add to the area.
- Integrate few purely pedestrian streets that are branched from the main street such that they include some restaurants, coffee shops, pubs and clubs.
- The slopes around the mall can help create a memorable place.

Design Concept:

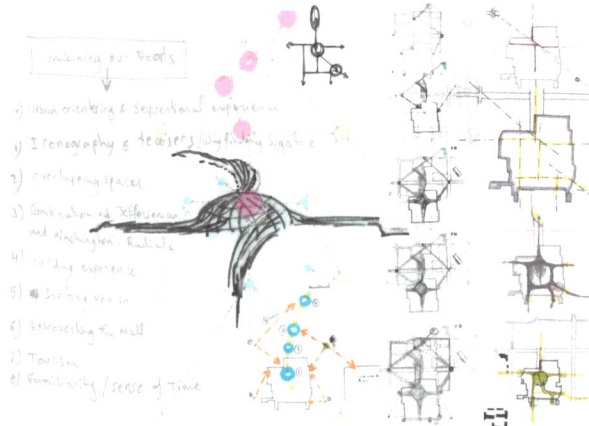

Fig.2: Sketches showing the combination of inspiring key words with some doodling.

Site Location:

Fig.1: A diagram showing the stretch of area thought off.

The major idea is how to combine all these pearls in an uninterrupted way and think about the experience from a pedestrian's point of view.

Fig.3: showing the existing conditions.

A fine pattern is clear but what is lacking is a sense of life, vibrancy, connectivity and maybe sensuality. Trying to integrate the new design, an architectural icon into the existing conditions.

Haneen Abdul Samad

Fig.4: A conceptual future vision.

An intention to give the center of the mall the major role of holding the area together. Extrovert the mall and reflect all the internal circulation pathways on the roof as skylights and all the way to the street level forming teasers and points of attraction.

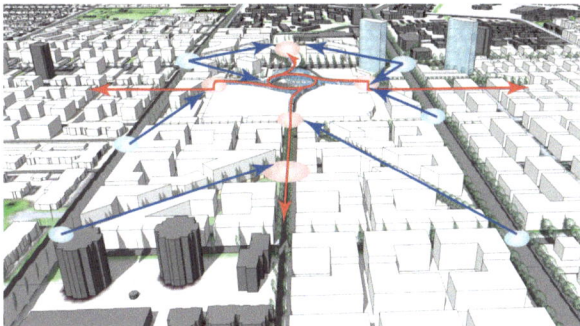

Fig.5: Urban Orienteering and Sequential Experience

- Emphasize all the entrances with minimal changes on their locations. (as shown in Fig.5 in light red)
- Find the major transit points, which can lead to major pedestrian tour locations. (as shown in Fig.5 in light blue)
- Link all these strategic points together. (as shown in Fig.5 in blue)
The result is a glimpse of a square and radial grid as shown in this sketch.

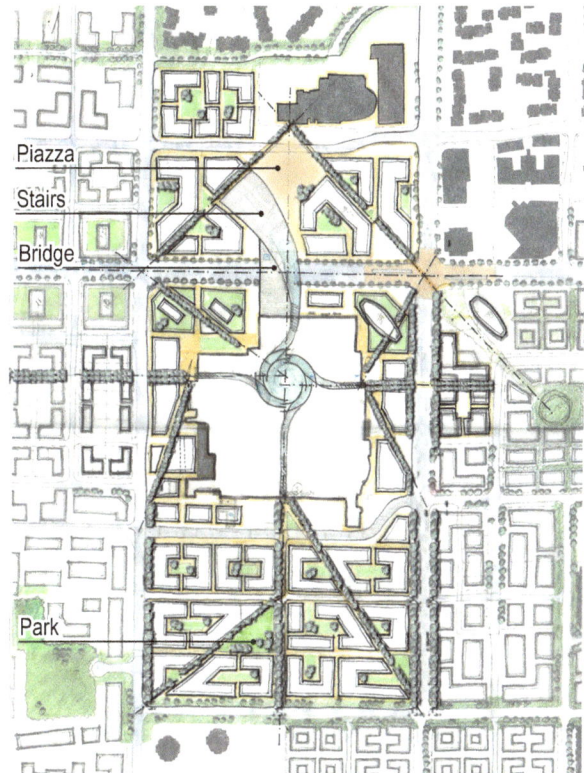

Piazza
Stairs
Bridge

Park

Fig.6: A Glimpse of the Combination of Jeffersonian Grid & Washington's Radials.

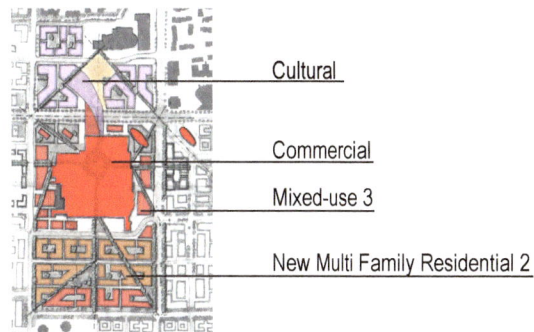

Cultural

Commercial

Mixed-use 3

New Multi Family Residential 2

Fig.6: Land use map

134

Concept

The vision of this proposal is to create Guildford as a new cultural and commercial centre of Surrey and build the old town center as the vibrant central heart of Guildford. The dashed line shows the study area and the chosen site is indicated in red.

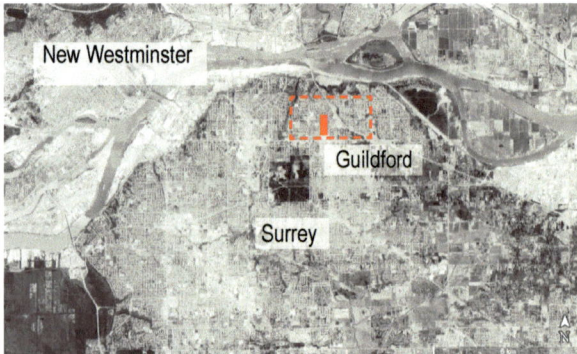

Figure 1: Site location. The site is located along the 104 transit corridor and between 150 St. and 152 St.

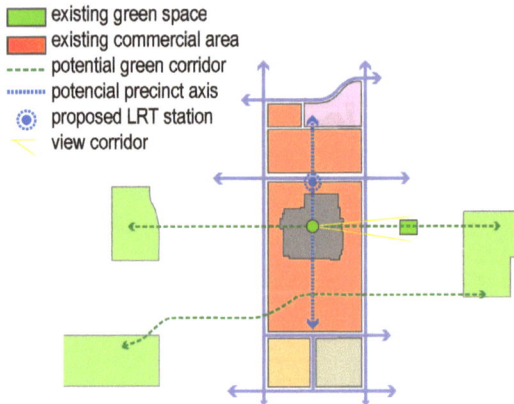

- existing green space
- existing commercial area
- potential green corridor
- potencial precinct axis
- proposed LRT station
- view corridor

Figure 2: Opportunities and existing land use

Based on the existing condition, there are some opportunities here. Firstly, there is a potential to propose a linkage between the two big green parks, creating an open corridor. Second, because in this scheme, the proposed future LRT station will be under the bridge, so considering the main circulation, there is a potential to create an axis connecting the north part to the south part. The horizontal corridor and

Proposed Land Use

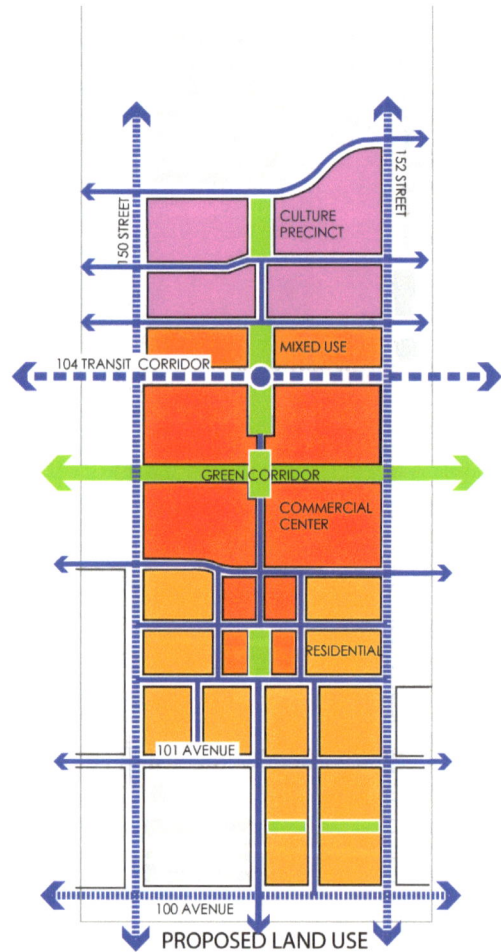

Figure 3: Proposed land use

vertical axis forms the main structure of the site.

The existing land use is taken up by a large part of commercial area. Besides, the site lacks an organized road network and this big-box fabric isn't a friendly and of walkable scale. Figure 3 shows the proposed land use, it will include cultural areas, commercial areas, mixed use areas, residential and green spaces.

Saki Xueqi Wu

Street Hierarchy

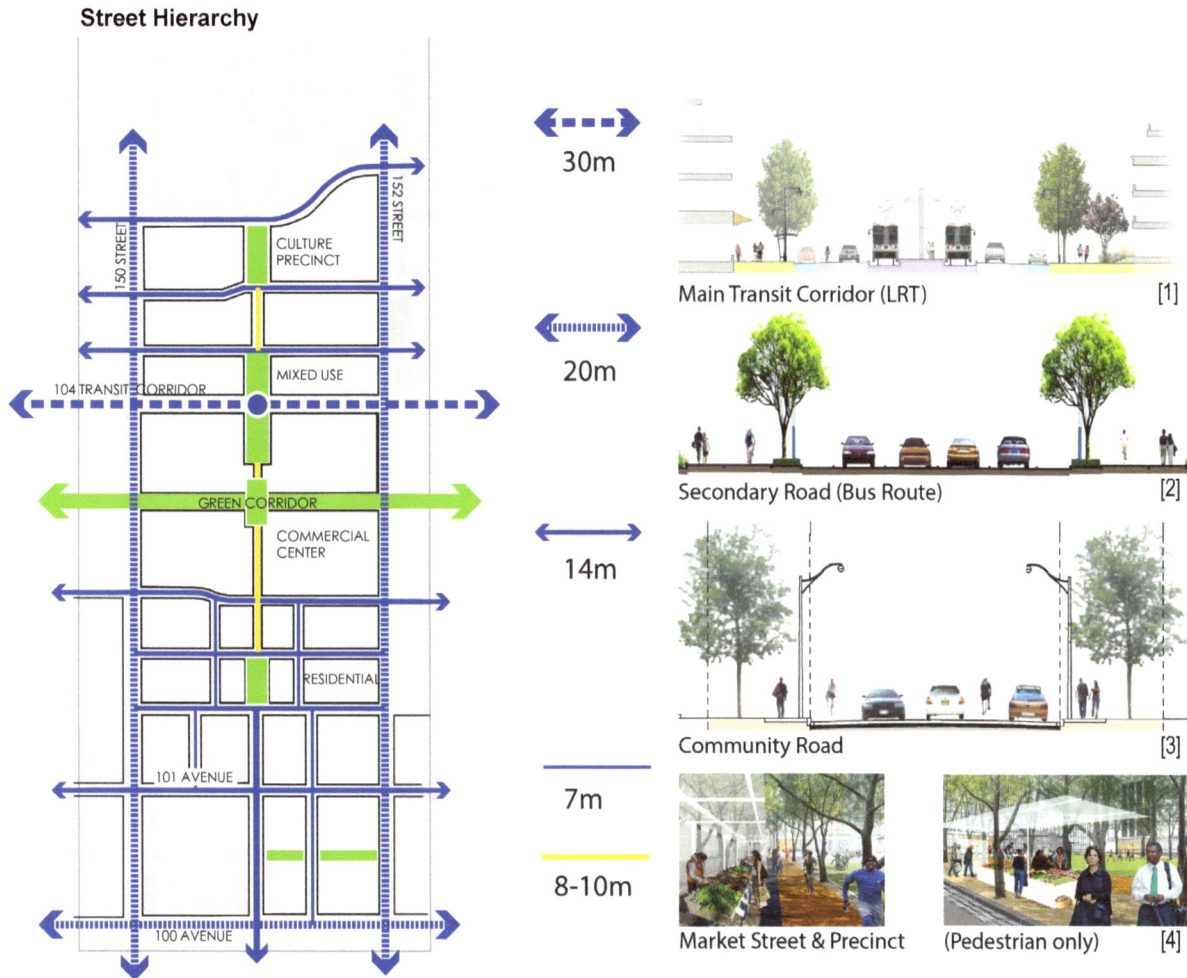

Figure 4: Street hierarchy, street's scale and selected section of each type of road.

Then, how can we achieve a vibrant heart? As we know, in addition to the buildings, the space between the buildings also plays an important role, which means a good street framework is required. That promotes diverse human activities on the streets. Figure 4 shows the street hierarchy in this area, which indicates five different types of streets. The main transit corridor will also be used by LRT. Second-ary roads (150 St. and 152 St.) will be used mainly as bus routes. Besides, there are community roads which are used by private cars. Furthermore, there are other two types of roads used only by pedes-trians that are cultural/commercial in nature and a market street. The market street exists inside the residential blocks, so that people could sell fruits and vegetables produced by urban agriculture.

136

Master Plan

Theater & Recreational Center

Library

Museum

Science&Technology Center
Bridge Park

Office Building

Mall
Central Courtyard

Entrance Plaza

Bar Street

South Park

Community Center

Typical Residential Area

[5]

[6]

[7]

[8]

MASTER PLAN

N

0   30   60(m)

Figure 5: Master plan and selected images of nodes.

Based on the previous analysis, Figure 5 shows the master plan. There will be a theatre and recreational centre, a library, a museum and a science and technology centre in the northside cultural precinct area. Mixed use forms will be built along the 104 St. main transit corridor which are eleven stories with two-storey podiums. The mall will be developed in several phases, surrounded by the new parking and retail buildings. The big box fabric will be broken down by some internal roads and a central court yard. A bar street will be a transition between the commercial area and the residential area. The south park will act as the end of the axis, used for outside concerts for people to get together. Besides, there is a community centre surrounded by the neighbourhoods and inside each neighbourhood blocks. We want to create diverse types of courtyards which will also be used for urban agriculture.

Saki Xueqi Wu

Perspective

Figure 6: 3D model of the Guildford Town Centre and surrounding areas.

The perspective view model shows the relationship between the site and the surrounding areas. The bridge will be the connection between the north and the south part through the roof gardens. Therefore, people can directly enter the mall from the second level.

Image sources:
[1].http://thetysonscorner.com/impacts-of-the-freddie-mac-wind-down-in-tysons/
[2].http://gardensofmylife.blogspot.ca/2011/12/walkside-cidade-feita-para-pessoas.html
[3].http://ottawa.ca/en/city-hall/planning-and-development/community-plans-and-design-guidelines/design-and-planning-0-1-7-4
[4].East Baltimore Urban Mixed-use District Design

[5].Southbank Cultural Precinct, Melbourne. http://www.fjmt.com.au/projects/projects_scpr.html
[6].Highline park in NYC. http://inhabitat.com/photos-iconic-high-line-park-in-nyc-opens-final-section-to-public/
[7].European city. http://www.big.dk/#projects-eur
[8].Southbank Cultural Precinct, Melbourne. http://www.fjmt.com.au/projects/projects_scpr.html

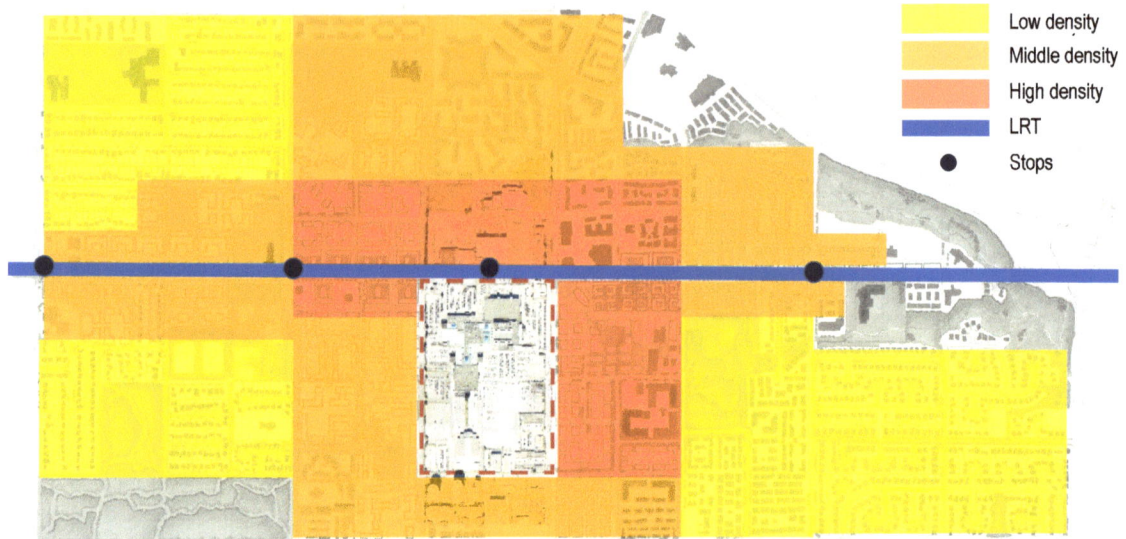

Figure 1 - The ascending trend of density

In the Guildford's proposed master plan, there is an ascending trend of density and building height from the Welcome Mats towards the Heart (Figure 1, refer to Managing Image and Scale Towards Strengthening Guildford's Identity). The site located in the urban centre of Guildford and is surrounded by 101 Ave., 104 Ave., 150 St., and 152 St. The site consists of several pieces of Guildford Town Centre Mall, parking area and some residential buildings.

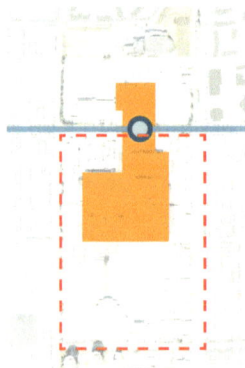

Figure 2: Mall and LRT stop

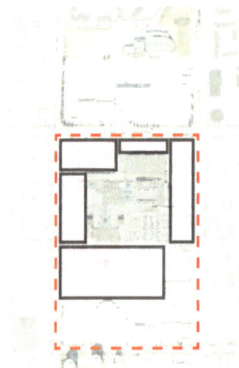

Figure 3: Parking lot

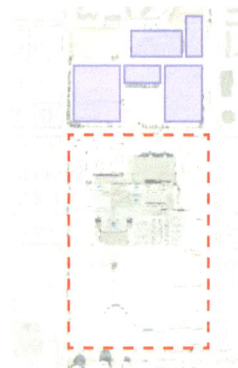

Figure 4: Culture centre

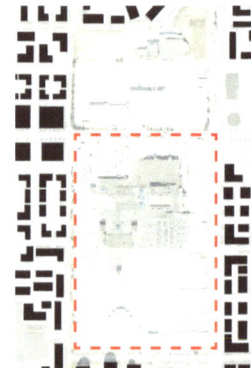

Figure 5: Residential Buildings

The LRT system is proposed in Guildford's master plan. A stop under the bridge will connects mall and northern extension area (Figure 2). Figure 3 illustrates the mall is surrounded by the parking lot.

The north side of the site has a proposed culture centre, which includes Guildford recreation centre and swimming pool (Figure 4). The location of the site is within a high density blocks (Figure 5).

139

👤 Siyuan Zhao

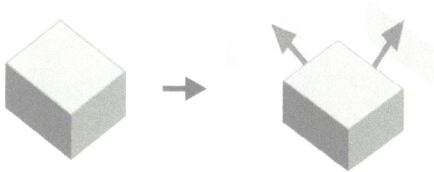

Figure 6: The Idea of open the mall

Guildford Town Centre Mall located in the central part of Guildford, is the second-largest shopping mall in British Columbia. For the vitality of the city, the mall just started a new redevelopment in 2010. Therefore, the mall has earned a value that must be retained. Considering the parking lot, the current mall is like a giant island isolated by the surroundings. The aim of design is to solve this problem, in terms of opening the mall and creating connections with neighbours.

There are three strategies across the study area that aims to open the mall: 1) from street to square, 2) place retail into mall extension, 3) bridge the gaps. The details of these strategies can be seen in the diagrams and images below.

Figure 7: Guildford Town Centre   Source:Dareell

Redesigning the boring street into a square, from mall facade to residential building facade, could change boundary of the mall into a sharing space. The activities, such as the outdoor Cafe, can be an attraction for gathering people and to improve the quality of life. Placing retail into the mall extension is also a strategy to maximize the use of the mall and save valuable space. Transparent texture (Figure 8) can soften the concrete edge of the mall and create a more urban atmosphere. Bridging the gaps is also a way to enhance the connection with surroundings and increase the shopping environment.

From street to square   Place retail into mall extension   Bridge the gaps

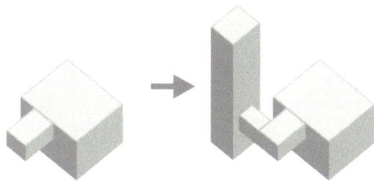

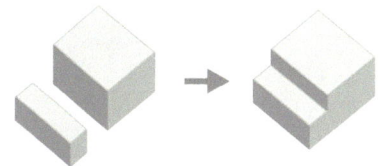

Figure 8: Three strategies of open the mall

Figure 9: Redesign of Stationsstraat
Source: Grontmij

Figure 10: Building on Water Street
Source: Googlemap

Figure 11: Bridge links Pacific Center and office tower. Source: Googlemap

The Master Plan

Figure 12: The main architectural structure in the northern part of the site is the Guildford Town Centre. It has retail as an extension and office tower linked with bridge. A proposed parkade replaces the current ground level parking lot. The connection between the commercial area and residential area is a street square. All the buildings in the southern part of site are all six stories high. Some buildings are pure residential. While some of them have commercial on the first floor and the top five floors are residential. The spaces between buildings is the green way and green space. The central green space emphasizes the central axis.

Siyuan Zhao

Figure 13: The land use of the site has a shift from pure commercial to pure residential.The buildings near mall and arterial roads have some retails on their first floor.

Figure 14: In order to increase the connectivity and walkability on the site, the roads are divided into several hierarchies. The streets serve pedestrians and cyclists, helping to create small-scale neighbourhoods.

A-A Section

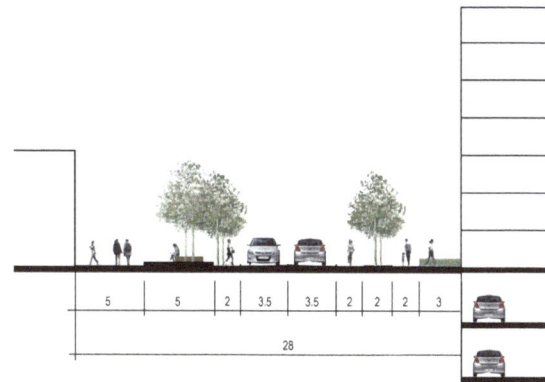

Figure 15: The street square is a combination of open space and typical street. It provides multiple functions and different space experience.

B-B Section

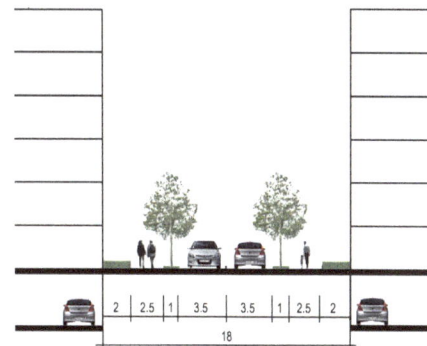

Figure 16: The section shows a typical street scale between residential buildings. The on-street parking is provided instead of underground parking.

STUDY AREA

LEGEND
- LRT Line & Stops
- Green Loop
- Study Area Boundary

The study area covers the mall site including the commercial block on the east and the plot of Guildford Recreation Center on the north. The scheme focuses on the larger area so the development on the surrounding parcels can interact with the regulation of the existing Guildford Mall. Covering both sides of 104 Ave., the site illustrates the potential to form additional north-south green connections to enhance the interconnectivity of the "Green Loop" for the overall master plan. The established local character as a retail hub is proposed to upgrade with greater diversity, reinforcing New Guildford as a welcoming place for everyone.

REGENERATION STRATEGIES OF THE GUILDFORD CENTER

GROUND FLOOR PLAN[1]   REGENERATION STRATEGIES

Sears (Two-story)

Guildford Town Center (Two-story)

Walmart +Parking (One-story)

LEGEND
- Regenerating & Adding Stories
- Infilling
- Demolishing & Constructing
- Disassembling Corridors

Various regeneration strategies are proposed for different parts of the suburban mall site, adapting it to the future urban context. The flat massing is proposed to be disassembled along the indoor pathways of the west wing and the northern demarcation of Walmart. These two disassembling corridors are proposed to be opened up to become in-block access streets. The major part of the mall will be reconstructed more intensively. New constructio infilling the existing parking lots will form coherent human-scale frontages for street retail and accommodate current tenements in the mall to ease the regeneration process.

Fig 1: Guildford Town Center Ground Floor Plan, Paul Hillsdon, Guildford Mall set for phase two of expansion, renewal. http://www.metro604.com/2011/07/26/guildford-mall-set-for-phase-two-of-expansion-renewal/, July 26, 2011.

Lilian Zhang

MASTER PLAN

DESIGN CONCEPTS

Two Retail Cores

Mini Green Loop

Five Integrated Function Clusters

The scheme reinforces the characteristics of the site as the retail landmark in the arrival community via diversifying the pedestrian experience for both the residents and the customers.

The retail circulation of the new urban mall will be stretched out from its daylighted east-west corridor to the east block. Linking the mall with the other key commercial element, the new proposed Guildford Marketplace. In addition, the indoor retail circulation of the mall and the street market along 153 St. will connect the 104 Ave. corridor and the neighbourhood on the south.

The mini "Green Loop", formed by the surface landscaped plazas and parks, roof gardens on the overbridge and the mall, and the boulevard of the neighbourhood, will integrate into an overall open space system for the New Guildford. This will thus enhance the interconnectivity of the "Green Loop" in the overall master plan.

The circulation network waved up by the commercial corridors and green spaces, will link the five clusters with various functions together and radiate the vibrancy to the surroundings.

FUNCTION LAYOUT & PHASING STRATEGY

Theater (Relocated)

Office / Hotel (Landmark Tower)

Retail Frontage + Parking in Podium + Apartments

New Guildford Center + Apartments (Urban Mall Complex)

Townhouse + Apartments Community

New Guildford Martket Place

Cinema

Mini Mall (Anchor Store) + Office

Retail & Office (Gateway Tower)

Market Street + Live & Work Units + Apartments

Multifamily House + Townhouse Community

| ■ Mall/Anchor Store | ■ Retail+Parking | ■ Townhouse/Apartment | ■ Multifamily House | ■ Workplace | ■ Retail/Market | ■ Culture+Recreational |

Phase I - 104 Ave. Frontage & Advanced construction for anchor stores relocation
- Blocks along the 104 Ave. corridor
- Mini mall, new Guildford Theatre, mixed use on Sears site

Phase II - Surrounding Construction I
- Gateway tower
- Regenerating Sears into cinema and mini mall
- Community along 152 St.

Phase III - Surrounding Construction II & Mall Regeneration I
- Guildford Market Village (reinforcing local characteristic)
- Community on the north part of the mall site (replacing Walmart)

- Complex block 1 on the northwest of the mall site (supplementing parking)

Phase IV - Mall Regeneration II (Adequate market target)
- Complex block 2 (regeneration of the west part of the mall)
- Expansion of the mall on the east with apartment tower
- Regenerating the bridge and the garden on its roof

Phase IV - Mall Regeneration III (Completion)
- Adequate market targetting the regeneration of the principal part of the mall (adding stories; daylighting east-west corridor; roof garden; townhouses and apartments on the top)
- Landmark tower

PHASE I

PHASE II

PHASE III

PHASE IV

PHASE V

145

Lilian Zhang

EAST BIRDVIEW

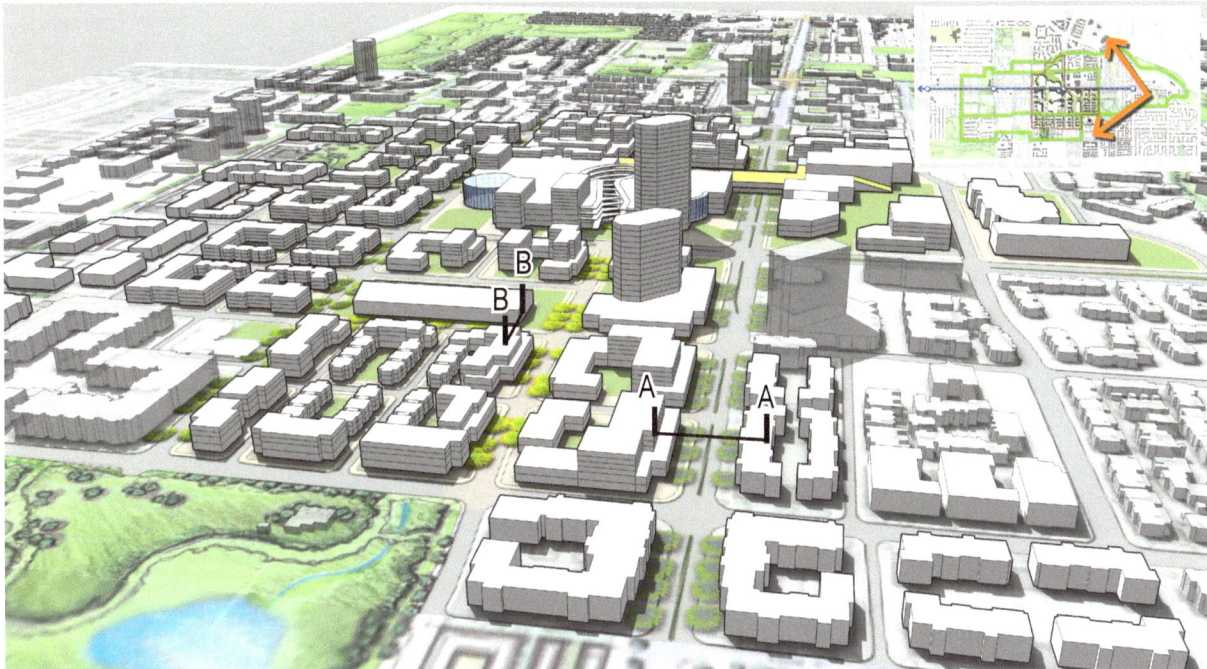

The New Guildford Mall integrates into the urban context creating a crucial node on the 104 Ave. corridor. The Guildford Market Village attracts visitors and customers by the unique retail stores supported by local residents with various cultural backgrounds. New towers and the overbridge highlights the site as the heart of this area. The site becomes a weekend destination with convenient rapid transit access.

With mixed use buildings with retail frontages along them, sidewalks become vibrant public spaces accommodating not only pedestrian flows but also various social activities.

SECTION A-A (104 Ave.)

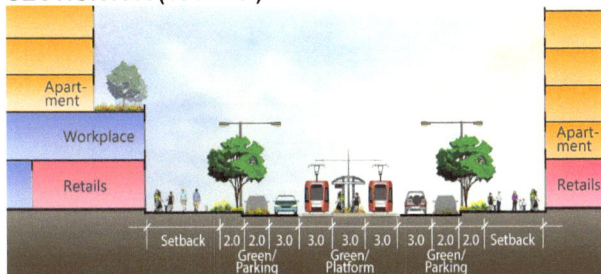

SECTION B-B (153 St.)

Cultured Pearls The North Side

This design embodies the central idea underlying the Guildford Neighbourhood Master Plan developed by a group of University of British Columbia urban design students "Transit-Oriented Culture". The principal aim of the Master Plan is to link together a series of six cultural centres or hubs located throughout Guildford area, which is a neighbourhood of Surrey, British Columbia. The hub featured in this plan, which is the most important of the six, is located adjacent the north side of Guildford Mall (fig. 1). It will include a new cultural complex consisting of a Cineplex movie theatre complex, a library, art school, museum, daycare centre, bowling alley, restaurant, coffee shop and other cultural and recreational amenities.

Figure1: Site location and The Cultured Pearls in master plan

The site of the hub is strategically located near a proposed LRT transit stop, Guildford Mall and a proposed greenway (fig. 2).

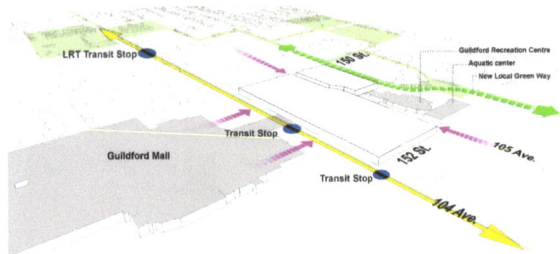

Figure2: Site Analysis

The structural forms are oriented in relation to five intersecting axes that constitute the principal pedestrian desire lines that provide access to the transit stops, mall and greenway (Fig. 3-4-5-6).

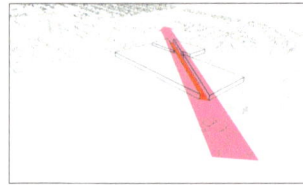

Figure3. Pedestrian desire line from the proposed LRT stop toward the proposed green way

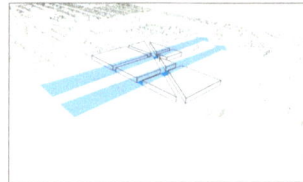

Figure4. Pedestrian desire line from the Guildford mall toward the proposed green way and existing recreational center.

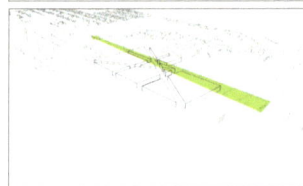

Figure5. Pedestrian desire line on the proposed green way.

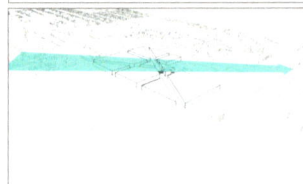

Figure6. Pedestrian desire line from the proposed LRT stop toward the proposed green way and existing recreational center.

After intersecting the lines, which can be considered corridors or axes, intersect each other, and in so doing dictate the shape of the structural forms (Fig. 7).

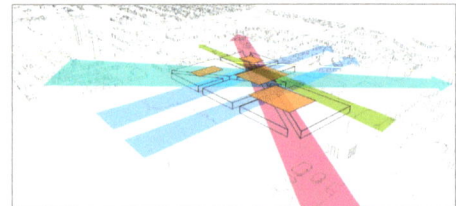

Figure7.

Nastaran E.Beigi

In addition, after intersecting, the axes open up a series of plazas, two smaller ones on the west side of the site and two major ones located, respectively, in the center and on the east side, which serve as an urban extension of the buildings that will house cultural amenities. This hierarchy of plazas will define the activities to be accommodated at this part of the site (Fig. 8).

Currently, there is a pedestrian bridge connecting Guildford Mall to the Sears department store located on the other side of 104 Ave. The new design calls for major modifications to the bridge so that it will serve as a greenway (Fig. 9).

The design aim is to add value to the hub and provide access to its inner spaces via the bridge and a series of green roofs, while at the same time transforming the east plaza, one of four located in the hub, and the adjacent staircase into a dynamic urban space (Fig. 10).

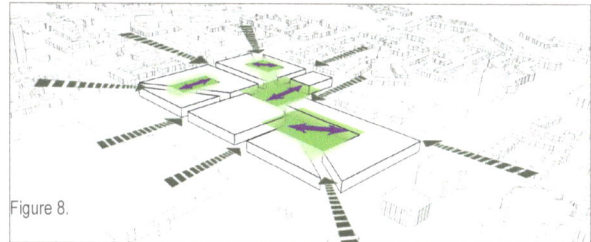

Figure 8.

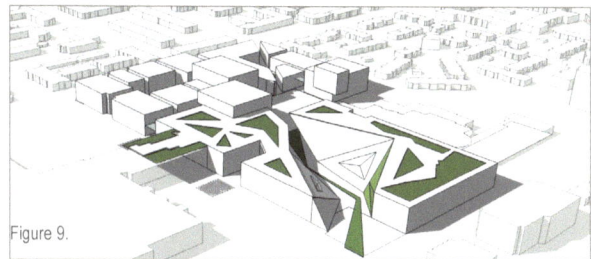

Figure 9.

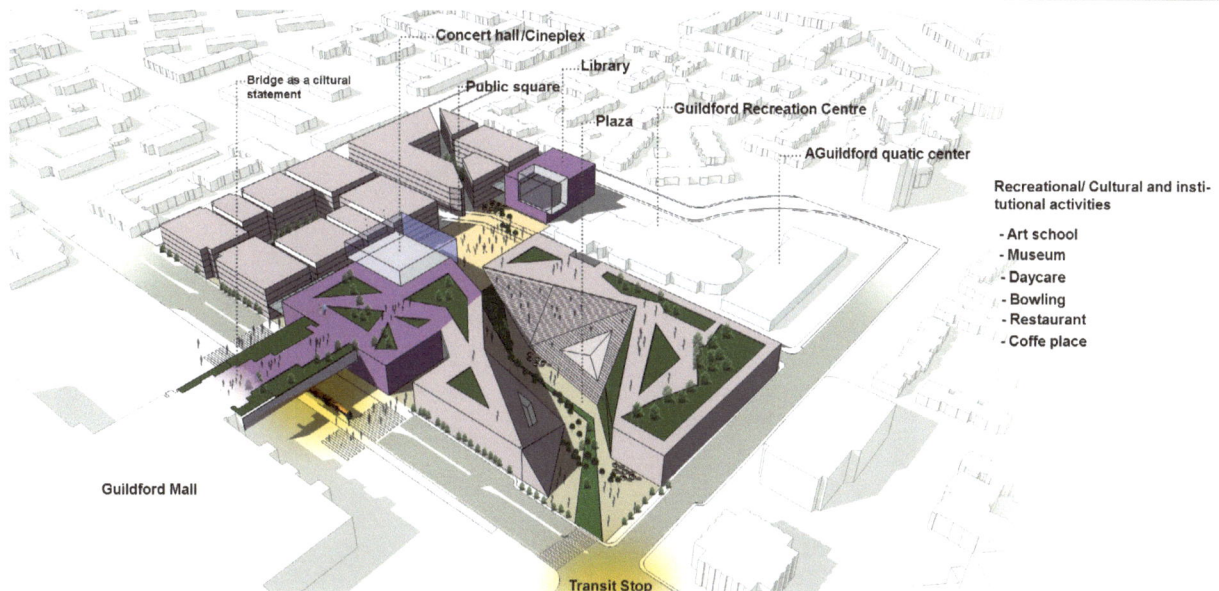

Figure 10. Site of the proposed Concert Hall and Cineplex, with the new library to be located along their axis. These two buildings constitute one corner of the plaza.

Resource: UBC Geography Information Center

Figure 11. The master plan illustrates that the public square is capable of accommodating daily users and large cultural events, while connecting with other plazas ans cultural centers.

The pedestrian bridge and staircase provide access to the green roofs and they also serve as cultural statements. Rome's Spanish steps provide a compelling example of what is intended here (Fig. 11-12-13).

Figure 14. Entrance to the east side plaza

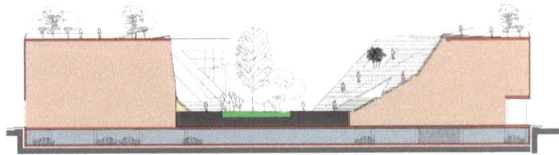

Figure 12. Section from the east side plaza and staircase.

Figure 13. Case study of Rome's Spanish steps which serves as a cultural statement.

Figure 15. A staircase for seating visitors attending outdoor concerts and theatre events held in the east plaza.

Nastaran E.Beigi

In conclusion, the design is intended to create a transit-oriented cultural complex that provides a dynamic urban space in which people can move, communicate and enjoy themselves, and at the same time have their needs met.

Figure 16-17. The green roof serves as the rooftop plaza.

Figure 18. An inviting public space accessible from 104th Avenue (the east side plaza).

150

THE CONCEPT

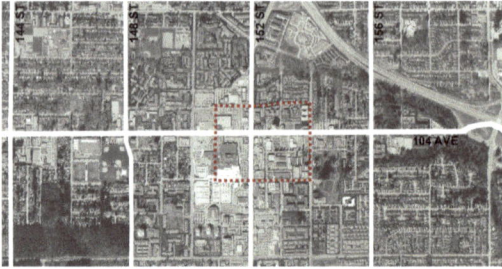

Figure 1: The site is located at the 104 Ave. and 152 St. In the future, there will be a proposed LRT Station at the cross

Figure 2: This figure shows the land use of this design. The red boxes are mixed use areas, and the yellow boxes are cultural areas

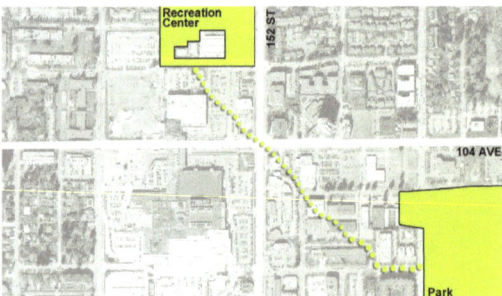

Figure 3: In order to link the cultural areas together, a green linkage will be used. At the same time, this linkage will link the recreation center and the park beside the site

THE MASTER PLAN

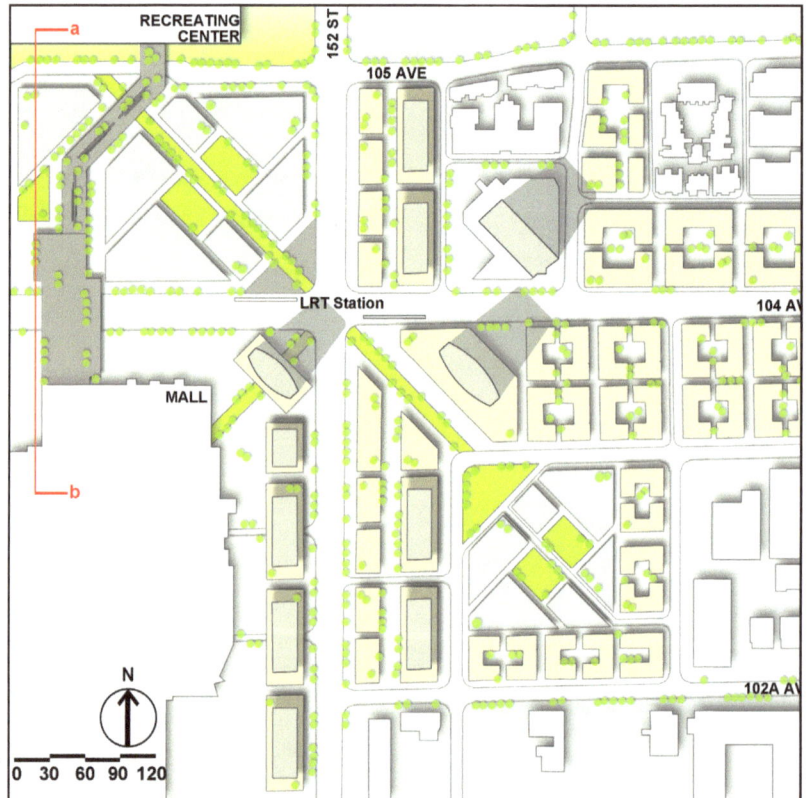

Figure 4: The Master Plan

Because of the Guildford Mall and the proposed LRT station, large amounts of human activities will happen in this area. According to the master plan (in the Figure 4), the green diagonal will be designed as a green way for pedestrians. On one side of the diagonal, the existing Sears building, which is beside the existing bridge, will be turned into a cultural park with some low-rise buildings and green open spaces. Meanwhile, the other side of the diagonal will be designed as a live-work area.

In order to increase the density, two towers will be added along 104 Ave. Another reason to put the towers there is to avoid their shadow falling on the buildings. The remaining part will be low-rise and mid-rise buildings. The buildings along the arterial roads will have mixed use functions, for instance, commercial use on the lower levels and residential use on the upper levels. The detail of building typology will be introduced in the third page.

Weicen Kate Wang

THE PERSPECTIVE

Figure 5 below is the overall perspective of this design. It shows that some low-rise buildings, as well as the podiums of the towers and mid-rise buildings, will have green roofs. The New York Highline will be used as an example for the bridge extension design. The Railspur Alley in the Granville Island of Vancouver will be the model for the pedestrian area design.

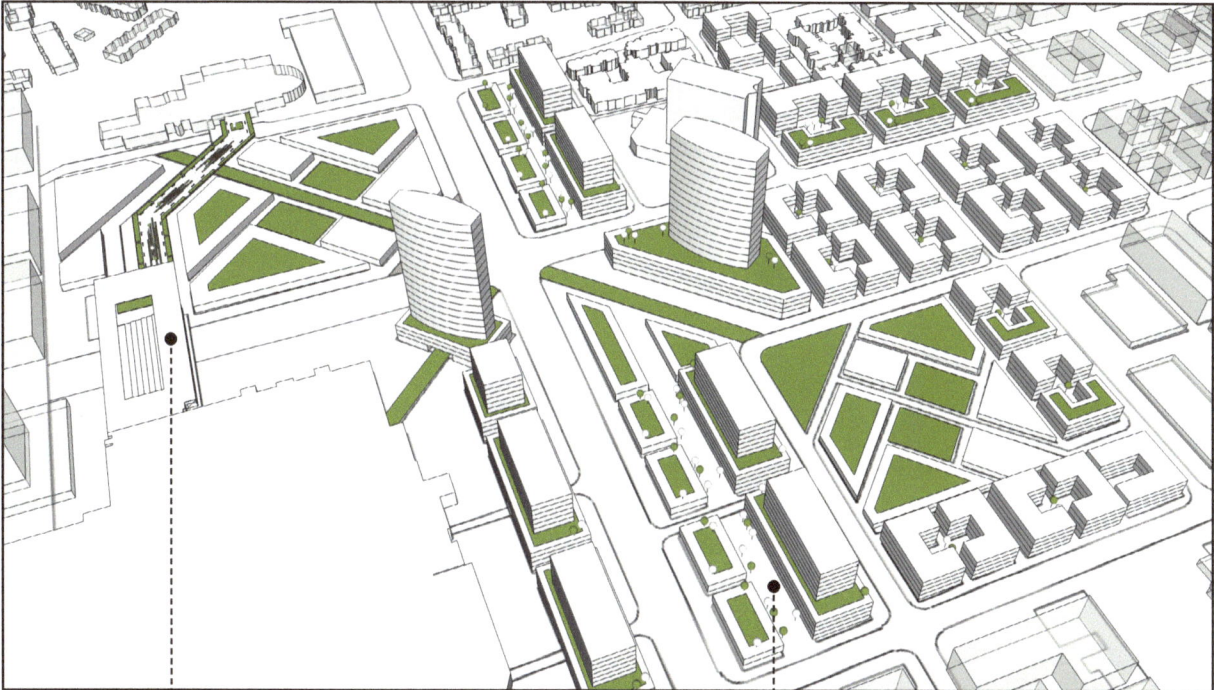

Figure 5 : The Overall Perspective

Source: http://lastblogonearth.com/2010/11/17/photos-of-new-yorks-high-line-park/

Figure 6 : The example of the bridge

Source: http://peitolake.blogspot.ca/2014/01/granville-island-vancouver.html

Figure 7: The example of the pedestrian area

152

THE BUILDING TYPOLOGY

TYPOLOGY 1

This type is located along the Guildford Mall. It is mid-rise building. The first floor will be used for retail. The lower levels will be the parking area, while the upper floors will be residential use. In order to let people easily access the mall, a path way will be left on the first floor.

TYPOLOGY 2

The second type is a mix of low-rise and mid-rise buildings. Functions such as retail, parking, and residential uses will be distributed in the buildings. A pedestrian area will be left for retail activities.

TYPOLOGY 3

The third type will be put along the 104 Ave. It consists of low-rise buildings. Because of the south-north orientation, the south part will be two floors lower than the north part, so that, the sunlight will not be blocked.

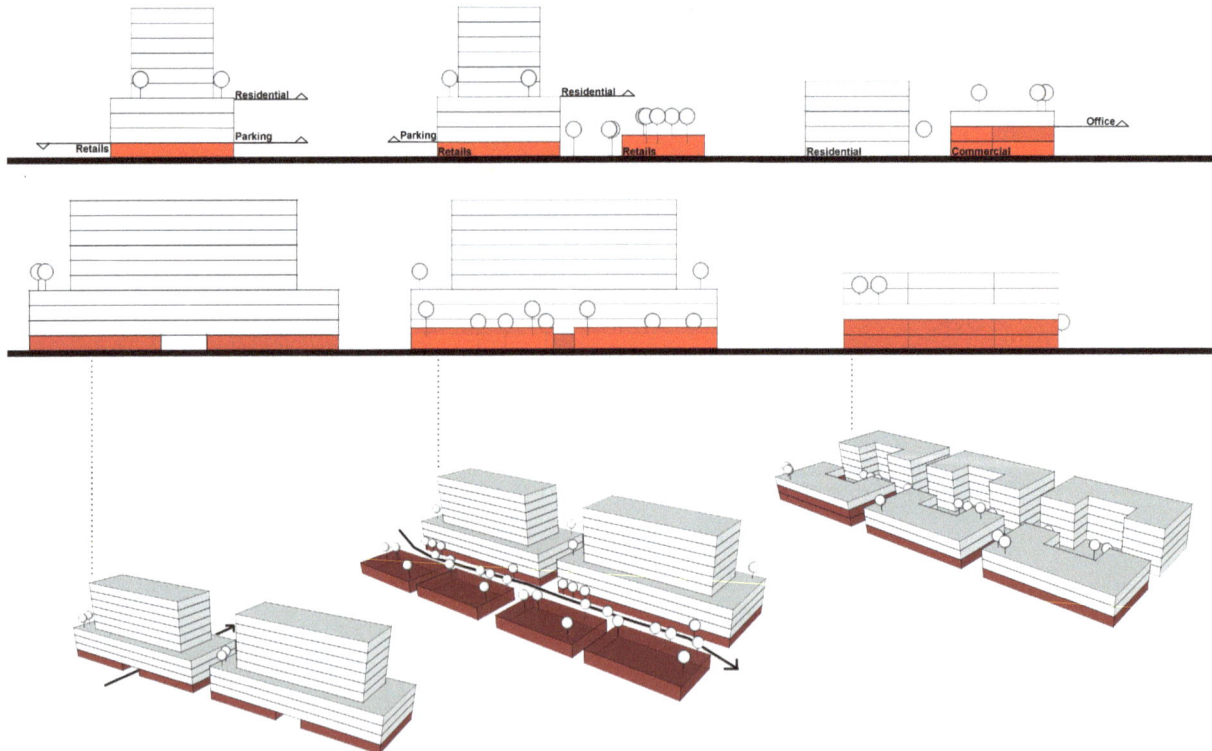

Figure 8 - Schematics of the building typologies in this design

Weicen Kate Wang

THE BRIDGE DESIGN

Source: http://en.wikipedia.org/wiki/High_Line_(New_York_City)
Figure 9: The example of the bridge's open space

Source:http://en.wikipedia.org/wiki/West_Side_Line_(NYCRR)
Figure 10: The example of the bridge's grey space

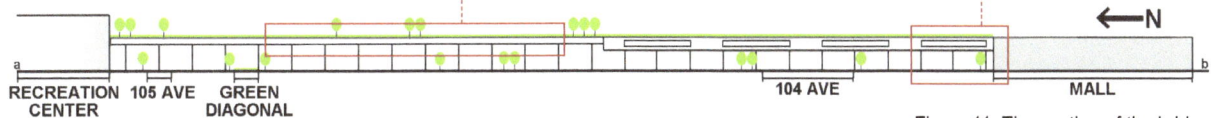

← N

a

RECREATION CENTER 105 AVE GREEN DIAGONAL 104 AVE MALL b

Figure 11: The section of the bridge

Figure 12: The perspective of the bridge's open space

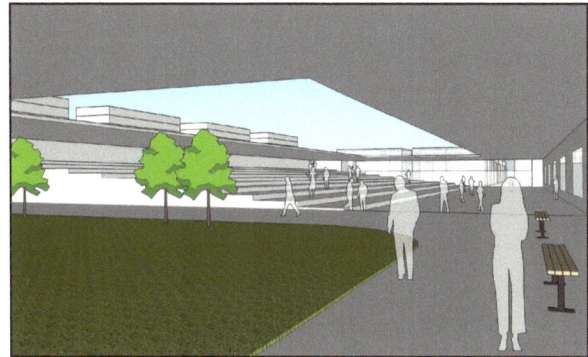
Figure 13: The perspective of the bridge's grey space area

Instead of demolishing the old bridge, this design will keep the existing part of the bridge, which is upon the 104 Ave. It will extend the bridge to become a linkage between the mall and the recreation center. Figure 11 above is the section of the bridge.

The existing part will be changed into a grey space with some stairs inside. It is shown in Figure 13. In this way, people who come out from the mall can take a rest on the stairs after shopping. The expansion part will be an open space, with some landscape designs on the top of it. It will directly stretch over the cultural park and connect to the recreation center. Figure 13 is a part of the landscape design of the bridge. On a sunny day, people can stay on the bridge, enjoy the beautiful view and sunshine.

154

Transect	Key locations	General Character	Frontage type	Type of Civic Space
T1		Reflective of local scale/ frontage with large courtyards and are intended to strengthen community and social exchange.	porches, shop fronts natural tree planting	courtyards and greens
T2		large landscape yards surrounding single family detached houses.	porches, fences natural tree planting	Greens and parks
T3		large landscape yards surrounding single family detached houses.	porches, shop fronts natural tree planting	Greens, parks and courtyards.
T4		preserved parks to facilitate neighborhood interactions.	Natural tree planting.	Park.
T5		residential midrise slab buildings atop a commercial podium which aids the new transit investment.	porches, shop fronts natural tree planting	Greens, parks and courtyards.
T6		The existing commercial hub of Guildford.	shops, front boulvards.	Squares, Greens and plazas.

T1 GENERAL URBAN ZONE | T2 SUB URBAN ZONE | T3 URBAN ZONE | T4 OPEN SPACES | T5 URBAN CORE | T6 COMMERC HEART

A transect was taken along the illustrated section represen Heights are purposefully neglected while tabulating the tran approach towards generalizing the urban character for the and diversifies harmoniously to result in the crescendo.
The character for Guildford is focused at 'Urban - Nature B nomics for the 104 Ave.

Yashas Hegde

T7 URBAN commercial CORE	**T8** URBAN ZONE	**T9** URBAN ZONE	**T 10** SUB URBAN CREEK

T7		buissiness hb that allows for increased job opportunities and adds economic vitality.	natural tree planting.	Courtyards, greens and squares.
T8		residential midrise slab buildings atop a commercial podium which aids the new transit investment.	natural tree planting.	Courtyards, greens and squares.
T9		Compact courtyard typologies which integrates with the surrounding open space.	natural tree planting.	Courtyards, greens and squares.
T 10		Natural landscape with creek.	natural tree planting.	Creek.

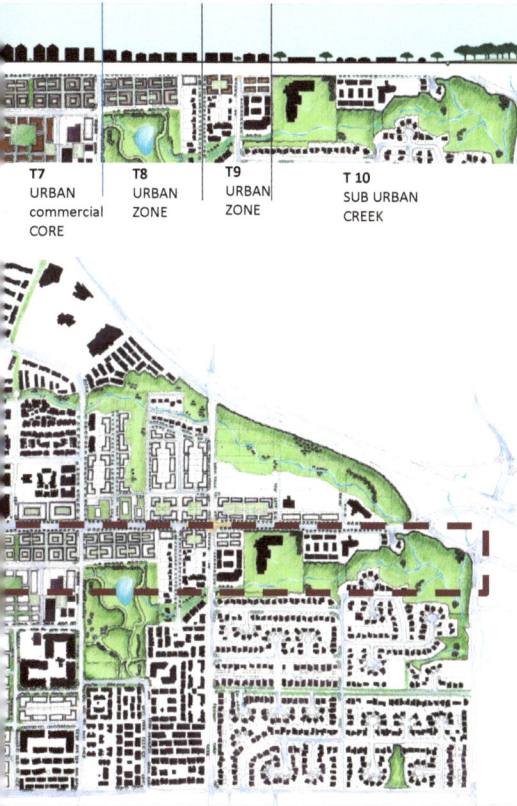

ove diagram to analyze the urban character of Guildford.
are regulated by zoning laws. This decision is a pragmatic
The building density is maximum at the commercial heart

re the flow of green spaces while respecting the land eco-

156

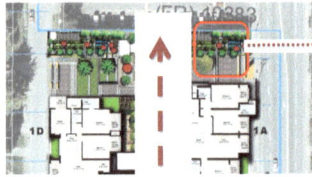

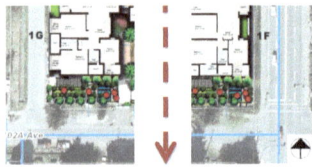

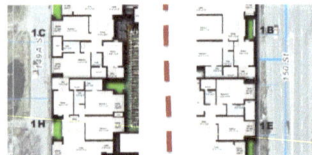

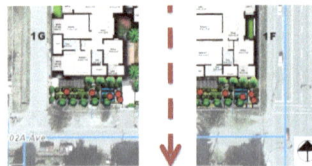

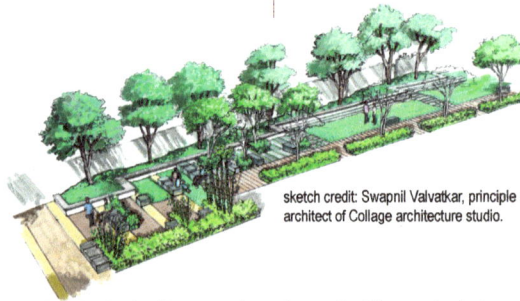

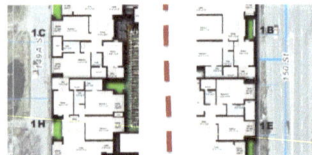

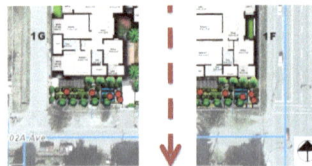

sketch credit: Swapnil Valvatkar, principle architect of Collage architecture studio.

sketch credit: Swapnil Valvatkar, princip architect of Collage architecture studio.

This sketch clealry illustrates how the pedestrian realm is harmoniously integrated with the built fabric. Swapnil's ideology in this sketch to create the mix of materials, hard scape and soft scape to create a vital pedestrian realm is integral to the place making identitiy.

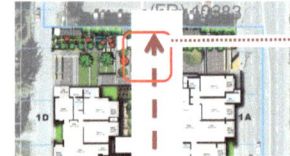

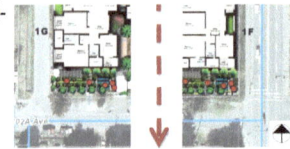

This sketch clealry illustra unifies street life and vitali creating a sense of place an animated people's plac of Swapnil's ideology, ama atmosphere.

sketch credit: Swapnil Valvatkar, principle architect of Collage architecture studio.

This sketch clealry illustrates how the pedestrian experience is split to break the monotony. This aids for a better pedestrian flow. The landscape is flown into the built urban fabric which not just ensures visual connectivity but also aids in connected green network which is essential for the a better ecosystem.

Yashas Hegde

"REVITALIZED URBAN CHARACTER"

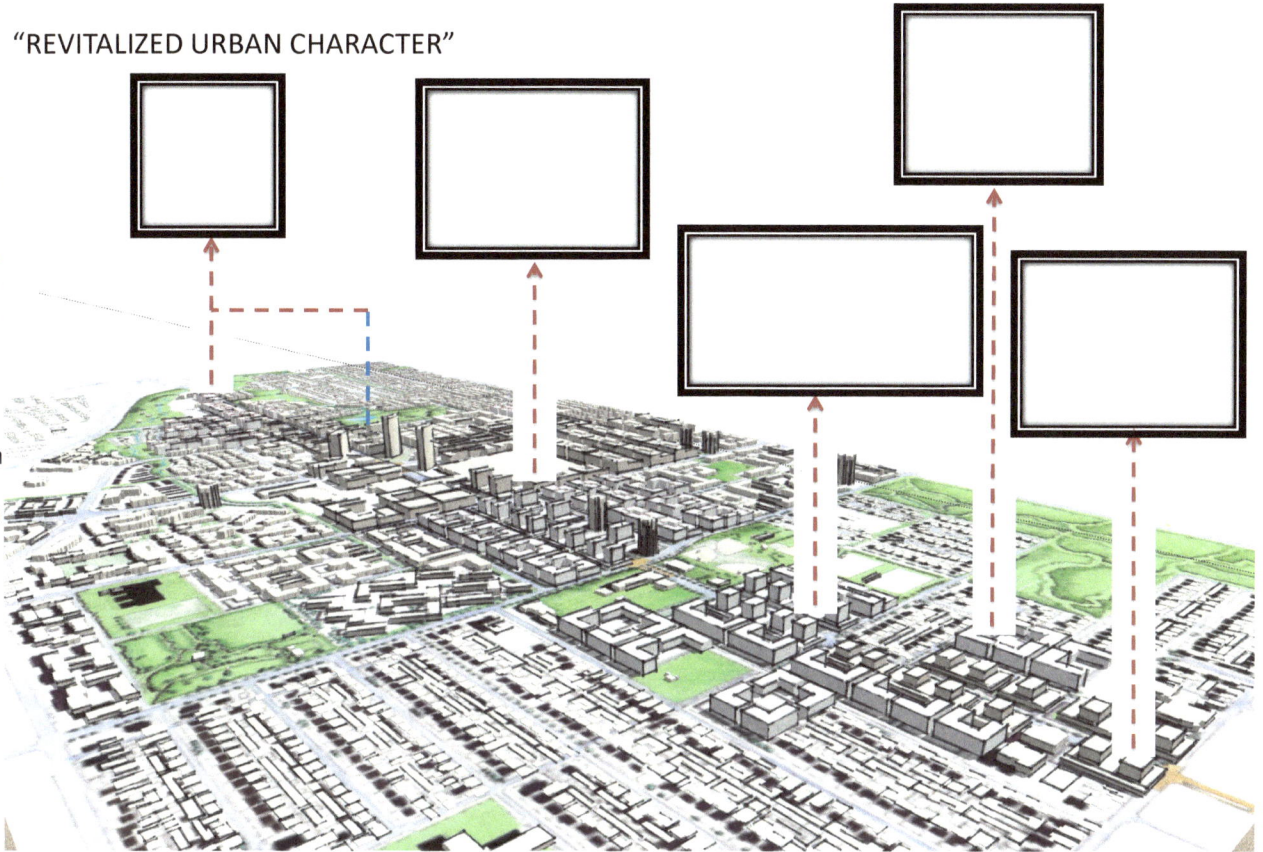

central public realm
fabric. This is vital in
the built fabric into
etches show a vision
create a unique

The design ideologies of my mentor in architecture, Mr. Swapnil Valvatkar, for new multi-family residential housing are represented in this study to represent how the flow of function should be arranged in order to attain a positive urban character. The duplicate trace off sketches of the original represents an ideology that integrates both the pedestrian streets and the built fabric . By maintaining this function order and by following the urban design principles, unique residential layouts that respect the needs of inner human desires, is amalgamated with land economics which can provide for unique multi-family residential ideologies. The form must flow the function preset of the site's urban character.

apnil Valvatkar, principle
ge architecture studio.

The Promise of Guildford:
Humaniẓ ng the 104 Avenue Corridor

The project aims at creating a humanized, walkable, vibrant and visually-engaging transit corridor at Guildford. In order to realize the goals, the spatial configuration of plots along 104 Ave. as well as the composition of rights-of-way were examined in terms of coherency, permeability, accessibility, connectivity, livability and vitality. Contributing more local and distinctive community expression and achieving more sustainable solutions, the project offers design solutions to create permeable blocks, well-connected green/open public spaces, coherent urban facades, visually-harmonized spatial configuration, human-scaled frontages, and a pedestrian-friendly transit-oriented corridor, all with regard to scale, nature and contextual considerations.

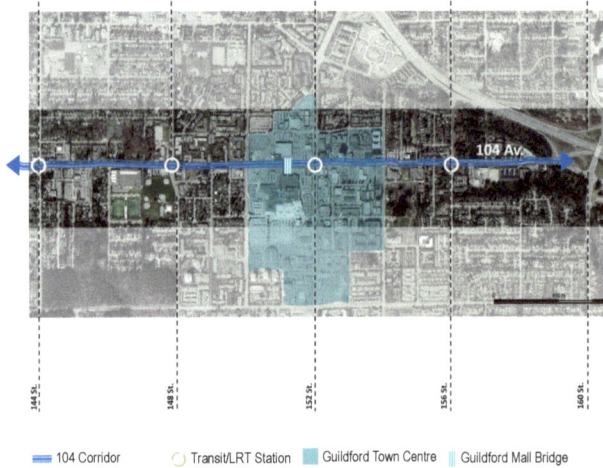

Transparent Facades and Permeable Blocks/Lots

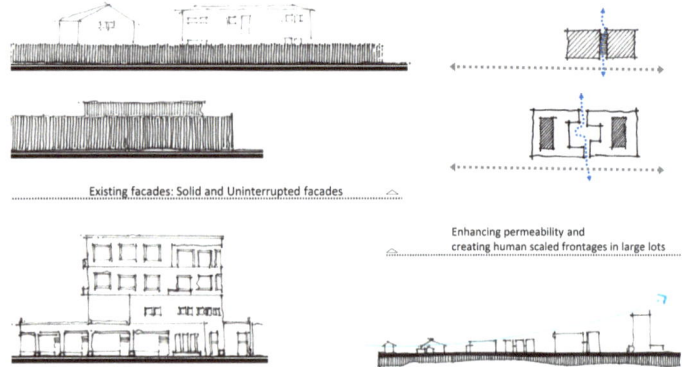

Existing facades: Solid and Uninterrupted facades

Enhancing permeability and creating human scaled frontages in large lots

Private land and pedestrian ᴢ ne

- 104 Corridor
- Transit/LRT Station
- Guildford Town Centre
- Guildford Mall Bridge

Building's Orientation (Over podium)
Minimizing shading impacts on public realm

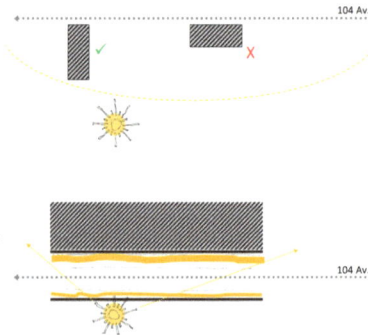

104 Av.

Northern sidewalk (Zone A:Mall): 10 m.
Southern sidewalk (Zone A: Mall): 6.5 m.

Northern sidewalk (General): 6.5 m.
Southern sidewalk (General): 4.5 m.

Alternatives
Desired

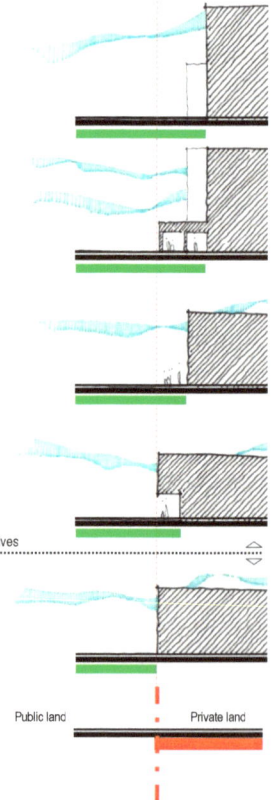

- Pedestrian zoning/Green zoning
- Private zone
- Property line

Public land   Private land

Livability/Vitality

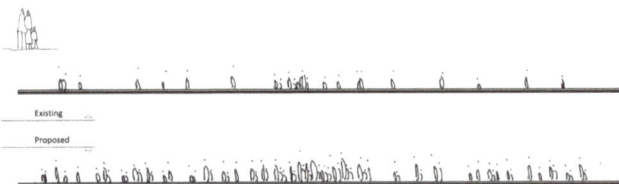

Existing

Proposed

Maryam Mahvash

Blocks and Permeability

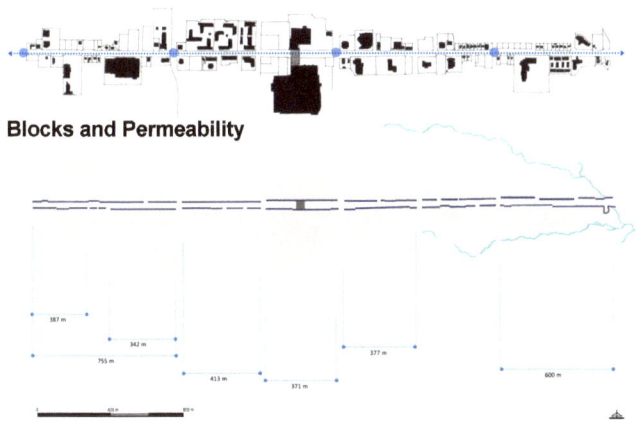

387 m
342 m
755 m
413 m
371 m
377 m
600 m

Public Realm

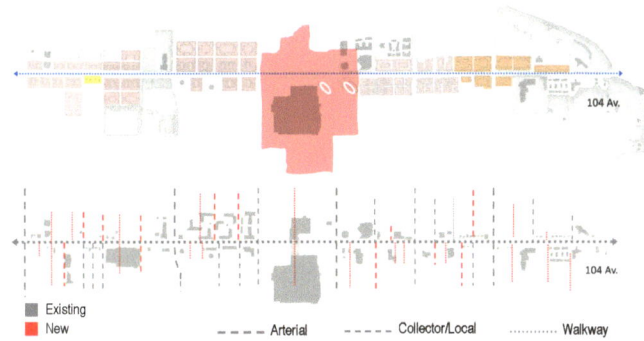

- ■ Park
- ■ Semi public green/open space
- ○ Plaza
- ● Iconic element
- ⌐ Public art

Accessibility

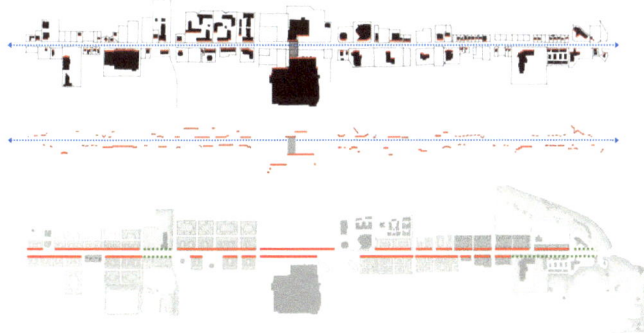

104 Av.

104 Av.

- ■ Existing
- ■ New

– – – Arterial   – – – Collector/Local   ·········· Walkway

Urban Transect

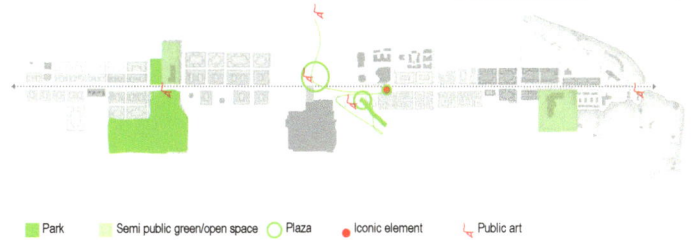

Transition: General Urban   Urban Core   Transition: General Urban

Less Urban   Medium Urban   Most Urban   Medium Urban   Less Urban

Urban Coherency

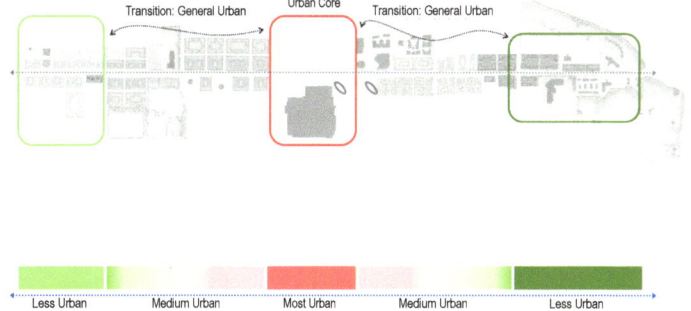

Vertical Rythm

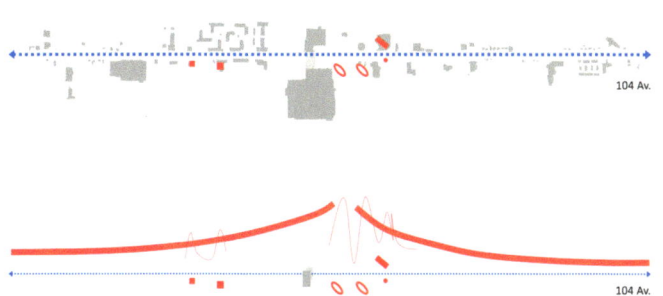

104 Av.

104 Av.

160

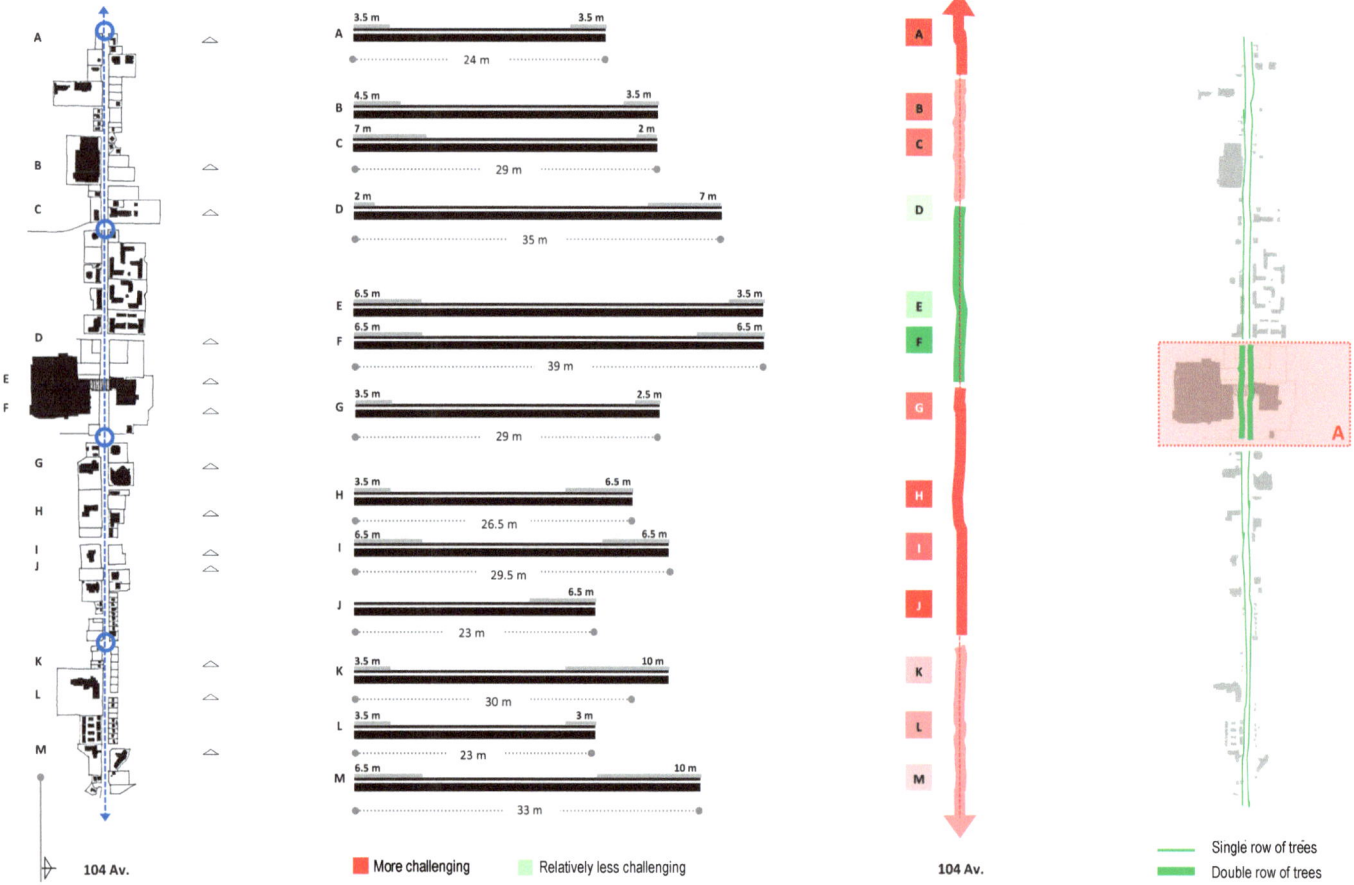

104 Av.

A	3.5 m — 24 m — 3.5 m
B	4.5 m — 3.5 m
C	7 m — 2 m — 29 m
D	2 m — 35 m — 7 m
E	6.5 m — 3.5 m
F	6.5 m — 39 m — 6.5 m
G	3.5 m — 29 m — 2.5 m
H	3.5 m — 26.5 m — 6.5 m
I	6.5 m — 29.5 m — 6.5 m
J	6.5 m — 23 m
K	3.5 m — 30 m — 10 m
L	3.5 m — 23 m — 3 m
M	6.5 m — 33 m — 10 m

■ More challenging   ■ Relatively less challenging

104 Av.

— Single row of trees
— Double row of trees

Dimensional variation along the corridor, illustrated by sections, is a major challenge. Resolving this major issue (dimensional constraints), the project has a specific focus on the design of rights-of-way and suggestions for both the configuration of mass and space and the determination of public (pedestrian) and private zones.

Proposed Section

Desired section

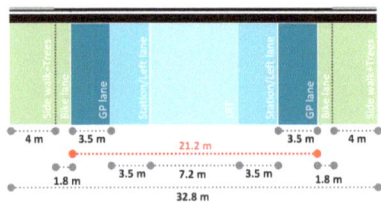

4 m — 3.5 m — 21.2 m — 3.5 m — 4 m
1.8 m — 3.5 m — 7.2 m — 3.5 m — 1.8 m
32.8 m

Proposed section (General)

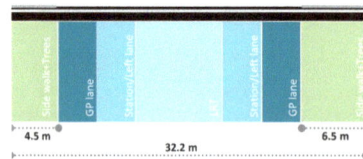

4.5 m — 32.2 m — 6.5 m

Proposed section (A: 150 St. - 152 St.)

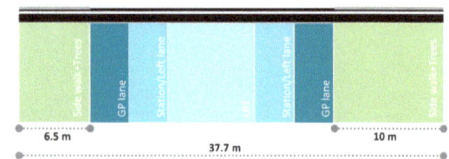

6.5 m — 37.7 m — 10 m

161

Proposed Railbed for LRT for Surrey

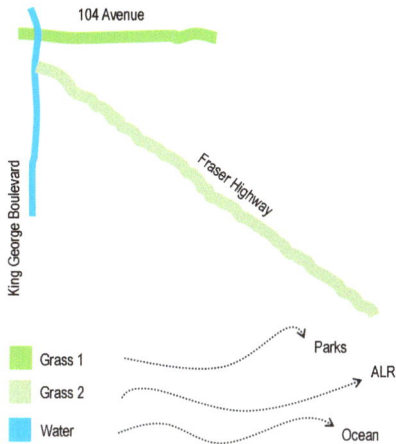

104 Avenue

Fraser Highway

King George Boulevard

Grass 1 ⟶ Parks

Grass 2 ⟶ ALR

Water ⟶ Ocean

Maryam Mahvash

Proposed Railbed for LRT for Guildford

Considering Light Rail Transit (LRT) on 104 Ave., a proposed grass rail bed perfectly offers a pleasant, humanized and environmentally-friendly corridor while the idea greatly meets some significant considerations of stormwater management. Based on this idea, a distinctive rail bed for two other LRT lines on King George Boulevard and Fraser Highway in Surrey are proposed.

View Down Proposed 104 Avenue corridor

Reference:
Kellett, Ronald and Girling, Cynthia. "Elements DB". elementsdb.sala.ubc.ca. 2010. WEB. November 29, 2014.

The Arrival City:

The Arrival City is a place in which the transition between rural to urban districts takes place. One of the main characteristics of the Arrival City is immigration. Considering Surrey's immigration rate, it is important to reflect the concept of the Arrival City. The Guildford area already serves as an arrival district in Surrey, therefore this character is maintained and improved in this design.

Economic Opportunities:

Most of the immigrant population that arrive in areas like Guildford are from lower income families, seeking for new economic opportunities. Hence this project mostly focuses on the economic aspects of the Arrival City, paying attention to opportunities such as small start-up businesses, retails and typologies including mixed use and live-work.

Location:

Accessibility is one of the main issues to be addressed regarding the implementation of small businesses and retails. Consequently, the first approach would be to locate these uses along the 104 Ave., since it is the major transit arterial. Although according to some features such as scale, width, traffic and noise along the 104 Ave., this design suggests to locate the site to the eastern side of the mall. This location is chosen based on the following reasons:

1- This site is fairly accessible since it is located between two major transit stops (152 St. and 154 St.).

2- It contains a narrower street network which creates a more human scale and therefore it is more likely to encourage pedestrian activity.

3- It is in close proximity to the mall and consequently encourages visitors.

4- The proposed commercial character of this area is different from the mall, thus it compliments it instead of competing with it.

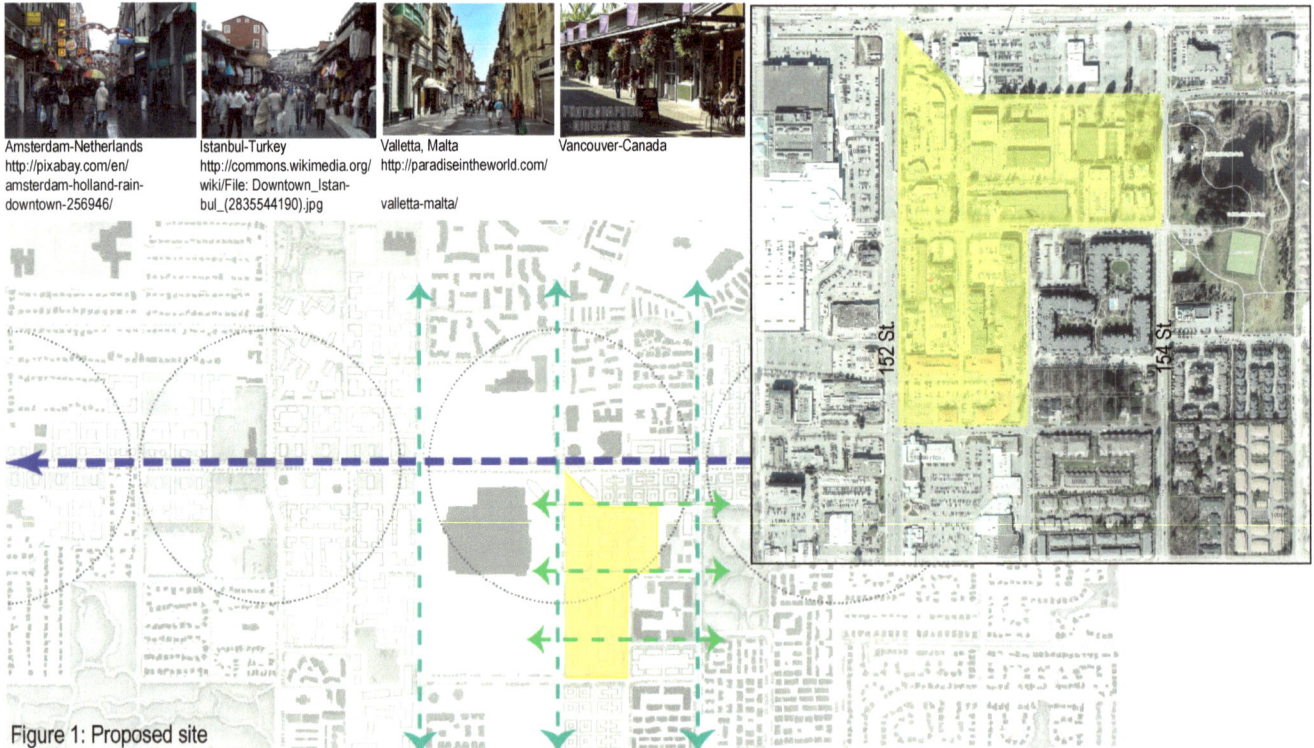

Amsterdam-Netherlands
http://pixabay.com/en/
amsterdam-holland-rain-
downtown-256946/

Istanbul-Turkey
http://commons.wikimedia.org/
wiki/File: Downtown_Istan-
bul_(2835544190).jpg

Valletta, Malta
http://paradiseintheworld.com/
valletta-malta/

Vancouver-Canada

Figure 1: Proposed site

Avishan Aghazadeh

Primary Design Objectives

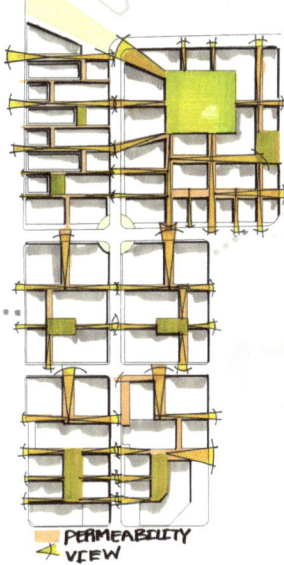

Figure 2: Permeability of the site which creates a sense of continuity within the fabric. Having various veiw corridors creates curiosity and therefore promotes walkability which is essential for start-up businesses.

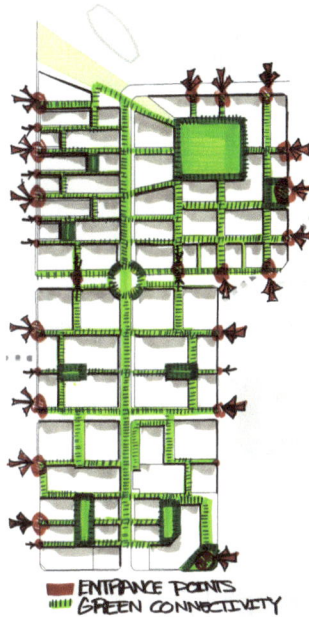

PERMEABILITY
VIEW

Figure 3: Many inviting entrance points. The green open spaces, connected through a network of green pathways, promote social communication and create an opportunity for cultural interaction.

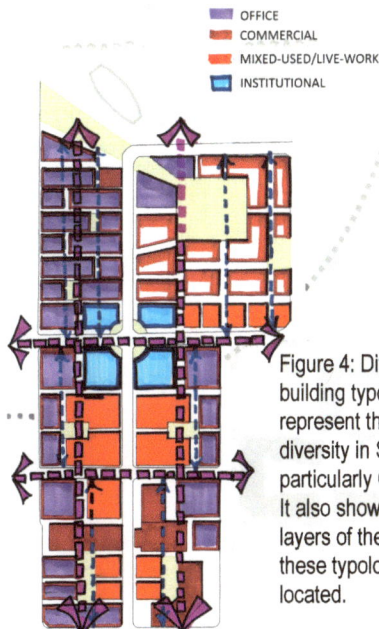

ENTRANCE POINTS
GREEN CONNECTIVITY

OFFICE
COMMERCIAL
MIXED-USED/LIVE-WORK
INSTITUTIONAL

Figure 4: Diversity in building typologies which represent the cultural diversity in Surrey and particularly Guildford. It also shows different layers of the fabric, where these typologies are located.

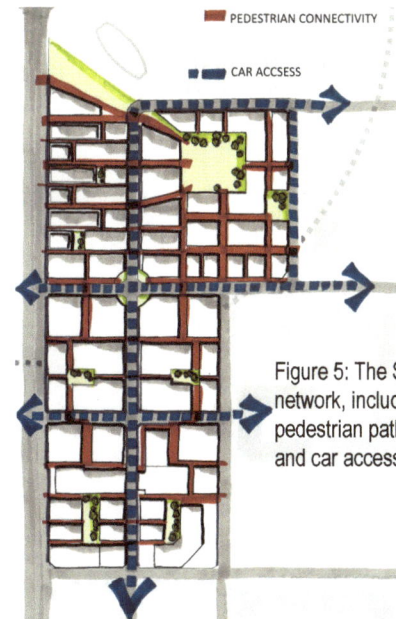

PEDESTRIAN CONNECTIVITY

CAR ACCSESS

Figure 5: The Street network, including pedestrian pathways and car access.

164

Figure 6: Site plan.
Buildings indicated in black are existing buildings. The footprint of these buildings will be preserved with an added density. The implementation of retail, offices, mixed use and live-work units can occur only through a rearrangement of the existing lots. The city may purchase some parts of the land with private ownerships, in order to implement the new design.

This design proposes implementation over different phases of time and through a series of new zoning policies. The new policies should address the need for having small businesses and therefore promote building typologies such as commercial, light industrial, mixed use and live-work.

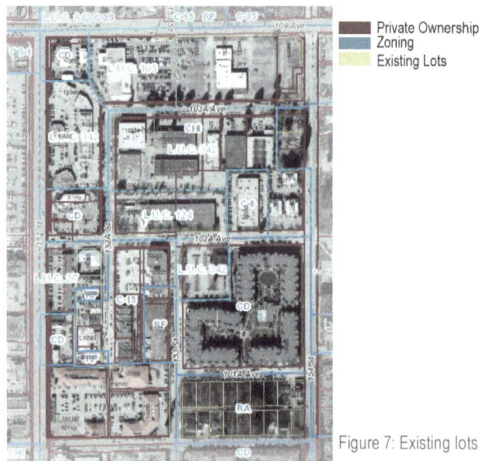
Private Ownership
Zoning
Existing Lots

Figure 7: Existing lots

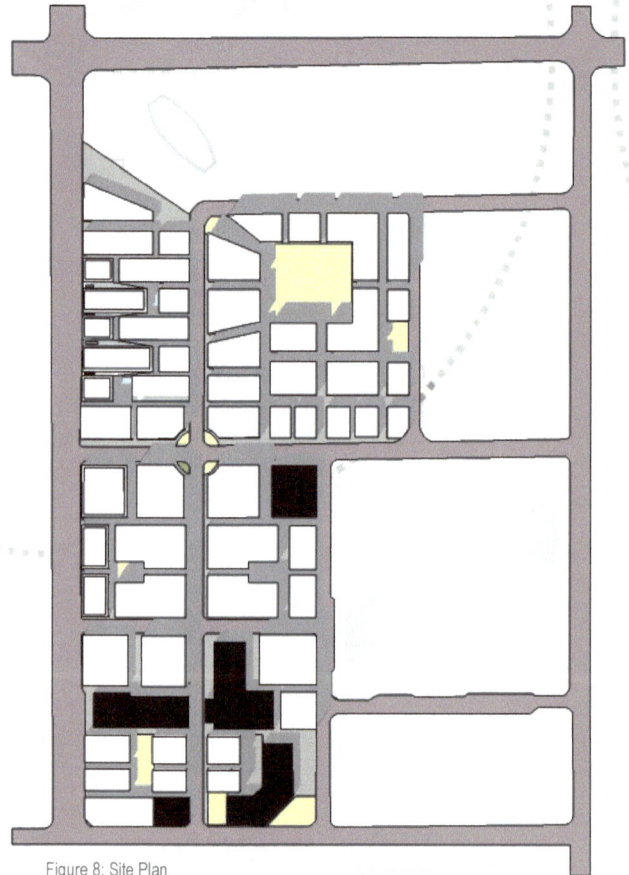
Figure 8: Site Plan

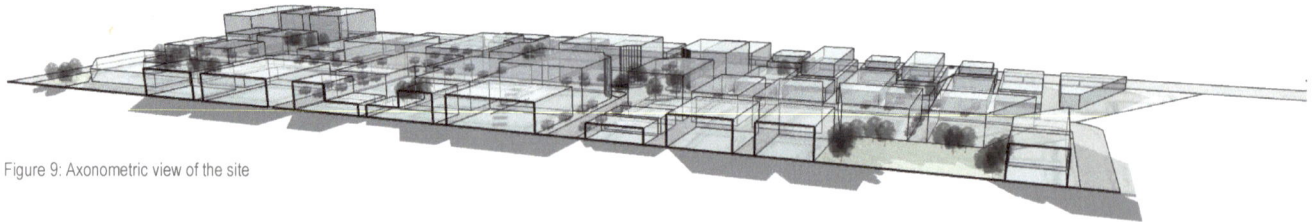
Figure 9: Axonometric view of the site

Figure 10: Site Section

Avishan Aghazadeh

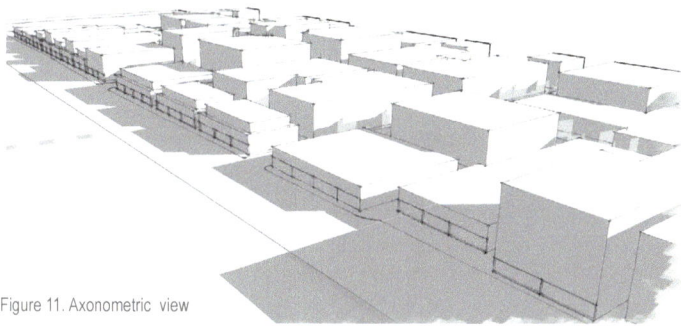
Figure 11. Axonometric view

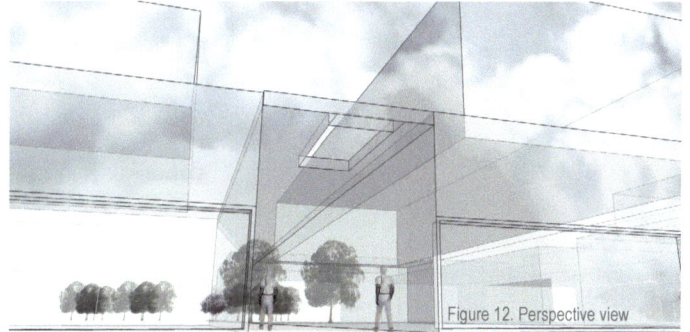
Figure 12. Perspective view

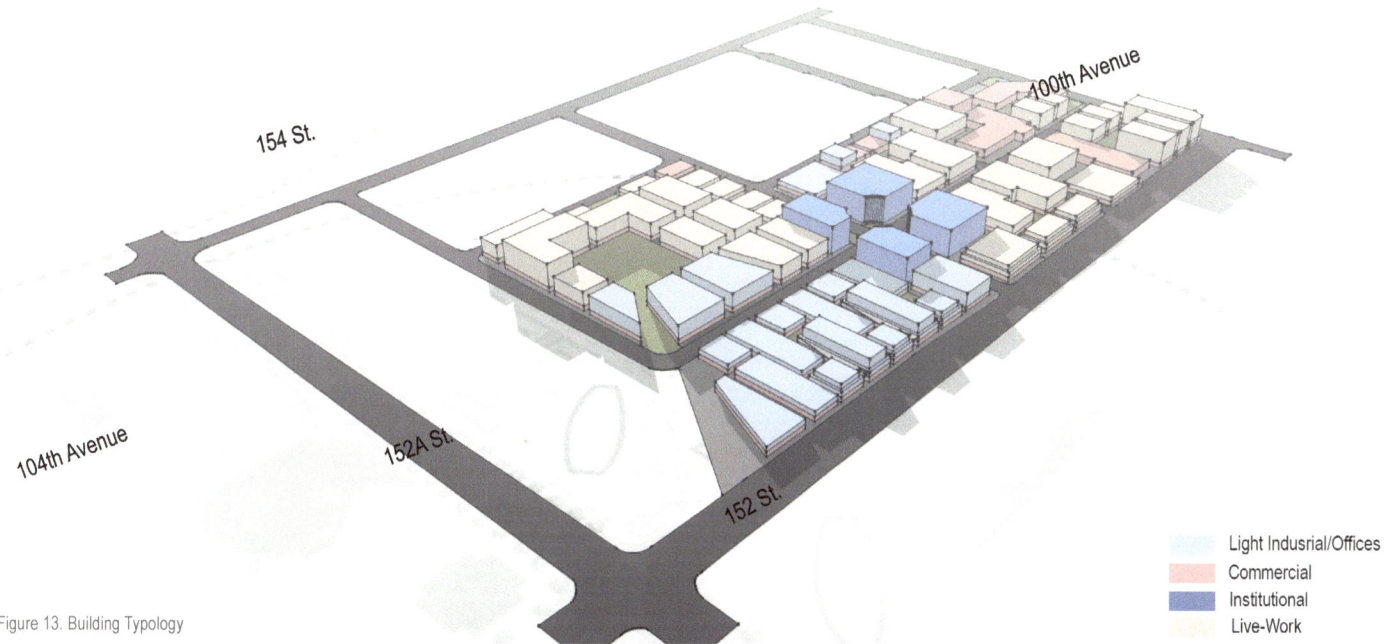

154 St.

100th Avenue

104th Avenue

152A St.

152 St.

Light Indusrial/Offices
Commercial
Institutional
Live-Work

Figure 13. Building Typology

Figure 1: Diversity in building typologies. The northern part of the152 St. corridor is the most accessible area to the mall and therefore mostly commercial and office buildings with very small frontages take place along this corridor. As we get closer to 100 Ave. and further away from the mall, mixed use buildings are proposed along the 152 St., and live-work units along the 152A St. At the north-eastern part of the site and along the 152A St., some office buildings as well as commercial and mixed-use buildings are proposed. This figure also indicates the two centres of this area:

1- The green open space at the north-eastern side: This is the main public space where cultural celebrations and festivals take place.
2- The institutional zone at the center of the site: This area contains four educational building such as a language school, business school and others.

166

The majority of Guildford's residents trace their origins to South Korea, China, Philipines and India (Fig.1). Major impediments experienced by new immigrants to the area include a lack of housing options and employment opportunities, steep language barriers and difficulties in subscribing to social services. Consequently, the current project addresses the twin problems of affordability and housing demand in Guildford for both, current residents and new arrivals.

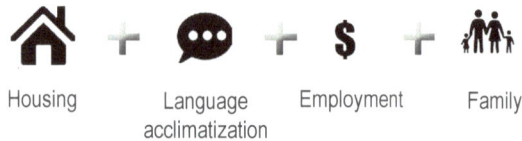

Housing + Language acclimatization + Employment + Family

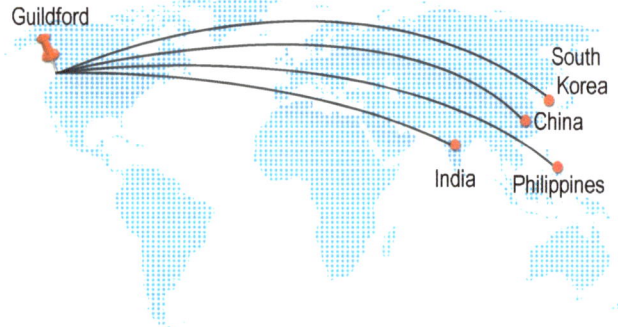

Fig.1: Map indicating flow of immigrants into Guildford

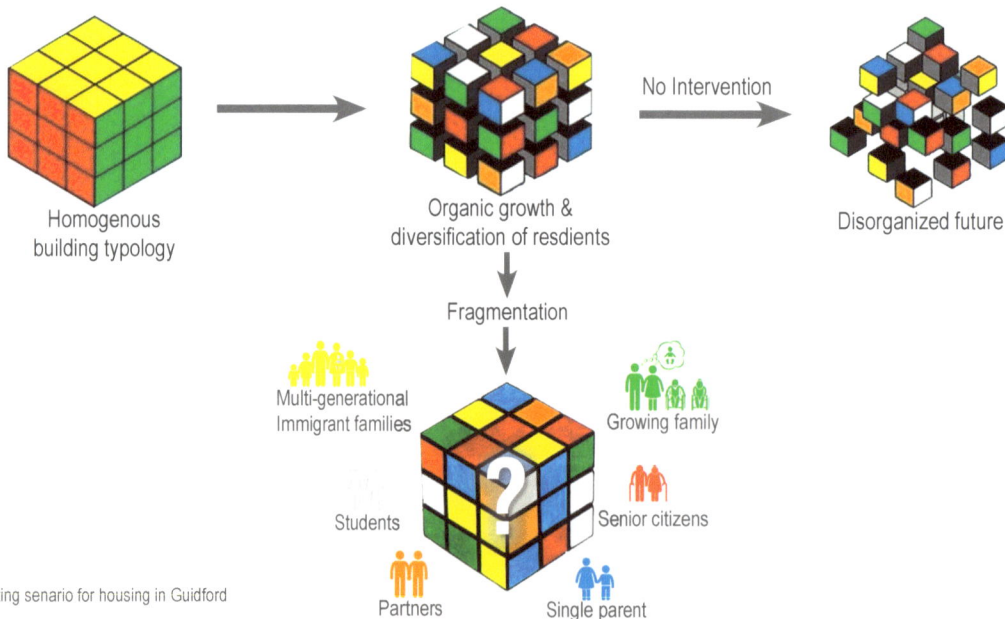

Homogenous building typology → Organic growth & diversification of resdients → No Intervention → Disorganized future

Fragmentation

Multi-generational Immigrant families

Growing family

Students

Senior citizens

Partners

Single parent

Fig.2: Existing senario for housing in Guidford

As depicted in schematic two, the current housing senario is mostly homogenous. However, as the economic prospects of the region have picked up in recent years, immigration into the area has increased significantly. The ensuing organic growth has resulted in fragmentation of building typologies. If revised housing and zoning policies are not implemented, the region could face a disorganized future. In view of this, the proposal aims to double housing capacity, release latent equity of the property, serve arrival city interests, and create value for home owners through rental stock. This will also house multi-generational families, senior citizens, single parents and other clientele.

Manali Yadav

Existing development:RS & RT (One and two family) Zoning ≤ 0.75 FSR

RM (Multi-family) CD (Comprehensive development) Zoning ≥ 1.6 FSR

Fig.3: The desired transition in housing typologies will be acheived by moving away from RS/RT zones to RM or related zoning

Incremental development

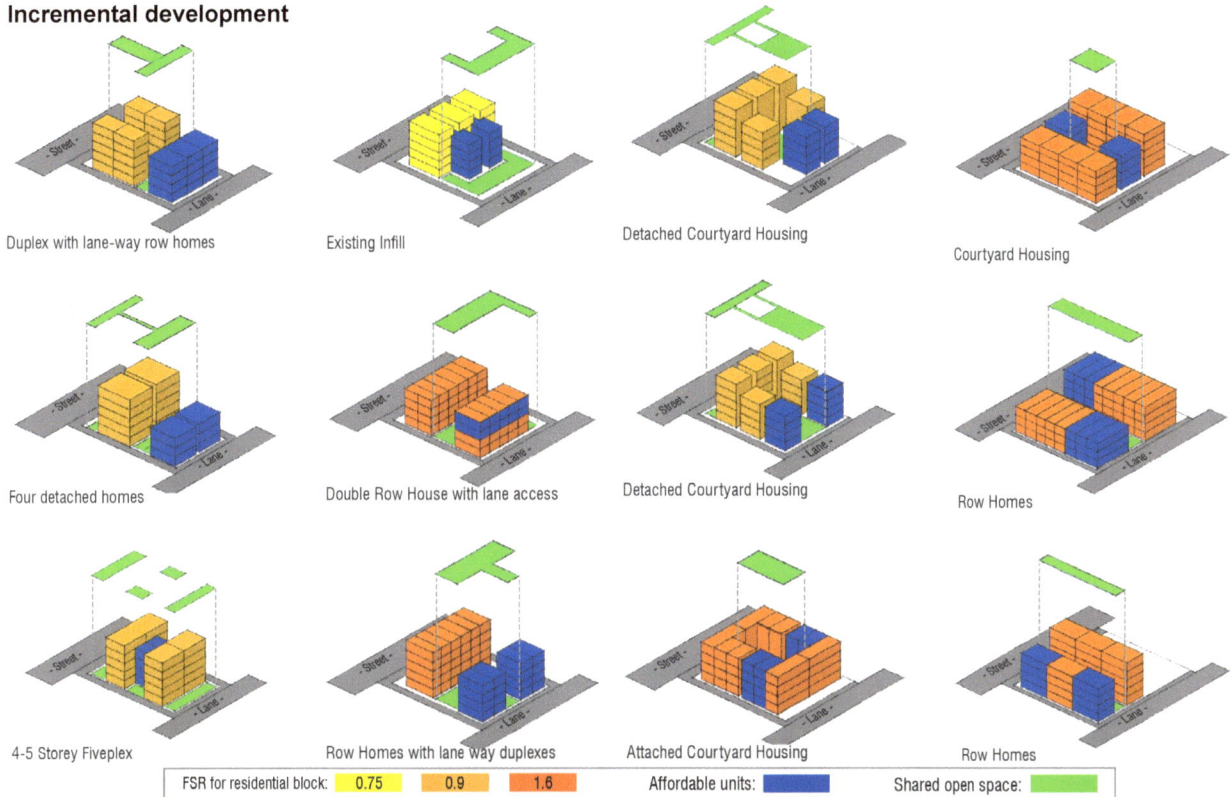

Duplex with lane-way row homes

Existing Infill

Detached Courtyard Housing

Courtyard Housing

Four detached homes

Double Row House with lane access

Detached Courtyard Housing

Row Homes

4-5 Storey Fiveplex

Row Homes with lane way duplexes

Attached Courtyard Housing

Row Homes

| FSR for residential block: | 0.75 | 0.9 | 1.6 | Affordable units: | | Shared open space: | |

By promoting small incremental development, as depicted in the schematic above, various building configurations such as row homes, infill houisng and courtyard buildings achieve housing densities between 0.75 to 1.6 FSR. These design layouts also include affordable rental units and shared open spaces that can by interpreted in many different ways. The spaces could be public, semi-private or private. The proposal encourages shared equity by offering minimal parcel assembly, thereby avoiding speculations about the land value.

168

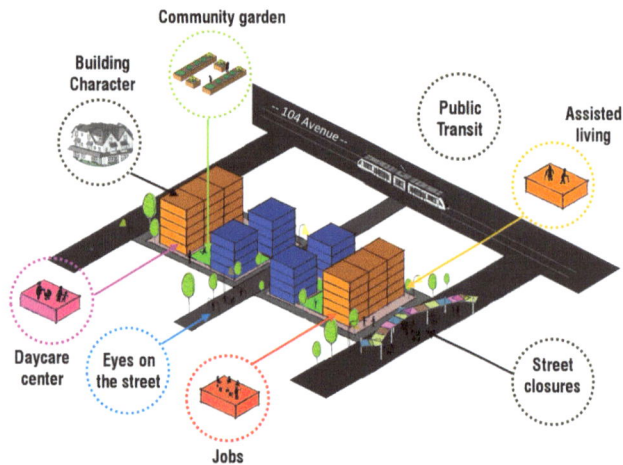

Fig.4: Necessary design strategies such as redesigning rental building character, locating institutional, recreational and social service agencies in close proximity are imperative for achieving a cohesive development

Schematic **4** provides the design template upon which successful integration of the new arrivals could be effectively achieved within the existing neighborhood. The development aims to encourage inclusionary zoning, which would free up as much as 25% of the housing in any newly redeveloped/rezoned residential area as rental stock. Such developments should be located close to public transit as it will be convinent to newly arrived families. The housing also offers live-work units, day-care centers, small community centers and assisted living for the senior citizens. Semi-private communal spaces are provisioned in order to promote community gardens, and street closures can encourage cultural and social events. As Guildford is one of the most diverse regions in Surrey, events such as religious congregations and food festivals will not only bring people closer to one another, but also help to gradually integrate them into society. The proposal caters to arrival city interests and emphasizes shared equity. Figs. 5, 6 & 7 depict the evolution of a city block over the course of 5, 10 and 15 years, respectively, based on developments anticipated through implementation of the deisgn stratgeies as listed in this proposal.

Suzie & Susan, owners of a single family homes, decide to consolidate their plots

Fig. 5: 5 Years later

Manali Yadav

Suzie & Susan enjoy their shared equity through rental earnings from the new court-yard housing that offers as much as 25% rental stock.

Fig. 6: 10 Years later

Guildford Mall

150th Street

104th Avenue

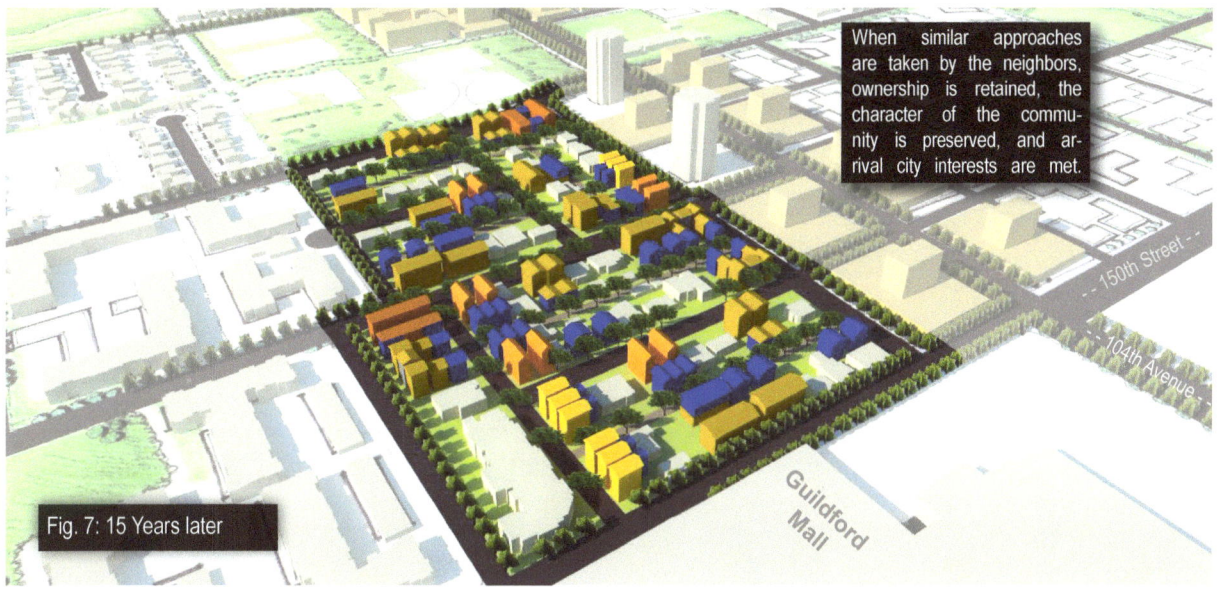

When similar approaches are taken by the neighbors, ownership is retained, the character of the community is preserved, and arrival city interests are met.

Fig. 7: 15 Years later

Guildford Mall

150th Street

104th Avenue

170

www.ingramcontent.com/pod-product-compliance
Lightning Source LLC
Chambersburg PA
CBHW041707210326
41598CB00007B/562